BEI GRIN MACHT SICH IHR WISSEN BEZAHLT

- Wir veröffentlichen Ihre Hausarbeit, Bachelor- und Masterarbeit

- Ihr eigenes eBook und Buch - weltweit in allen wichtigen Shops

- Verdienen Sie an jedem Verkauf

Jetzt bei www.GRIN.com hochladen und kostenlos publizieren

Bibliografische Information der Deutschen Nationalbibliothek:

Die Deutsche Bibliothek verzeichnet diese Publikation in der Deutschen National-
bibliografie; detaillierte bibliografische Daten sind im Internet über http://dnb.d-
nb.de/ abrufbar.

Dieses Werk sowie alle darin enthaltenen einzelnen Beiträge und Abbildungen
sind urheberrechtlich geschützt. Jede Verwertung, die nicht ausdrücklich vom
Urheberrechtsschutz zugelassen ist, bedarf der vorherigen Zustimmung des Verla-
ges. Das gilt insbesondere für Vervielfältigungen, Bearbeitungen, Übersetzungen,
Mikroverfilmungen, Auswertungen durch Datenbanken und für die Einspeicherung
und Verarbeitung in elektronische Systeme. Alle Rechte, auch die des auszugsweisen
Nachdrucks, der fotomechanischen Wiedergabe (einschließlich Mikrokopie) sowie
der Auswertung durch Datenbanken oder ähnliche Einrichtungen, vorbehalten.

Impressum:

Copyright © 2015 GRIN Verlag, Open Publishing GmbH
Druck und Bindung: Books on Demand GmbH, Norderstedt Germany
ISBN: 978-3-668-05847-7

Dieses Buch bei GRIN:

http://www.grin.com/de/e-book/307089/lie-symmetry-analysis-of-the-hopf-functional-
differential-equation

Daniel Janocha

Aus der Reihe: e-fellows.net stipendiaten-wissen

e-fellows.net (Hrsg.)

Band 1600

Lie symmetry analysis of the Hopf functional-differential equation

Lie-Symmetrieanalyse der Hopf-Funktionaldifferentialgleichung

GRIN Verlag

GRIN - Your knowledge has value

Der GRIN Verlag publiziert seit 1998 wissenschaftliche Arbeiten von Studenten, Hochschullehrern und anderen Akademikern als eBook und gedrucktes Buch. Die Verlagswebsite www.grin.com ist die ideale Plattform zur Veröffentlichung von Hausarbeiten, Abschlussarbeiten, wissenschaftlichen Aufsätzen, Dissertationen und Fachbüchern.

Besuchen Sie uns im Internet:

http://www.grin.com/

http://www.facebook.com/grincom

http://www.twitter.com/grin_com

Lie symmetry analysis of the Hopf functional-differential equation

Lie-Symmetrieanalyse der Hopf-Funktionaldifferentialgleichung
Master-Thesis von Daniel Janocha
Tag der Einreichung:

Fachbereich Maschinenbau,
Fachgebiet für Strömungsdynamik,
AG Turbulence theory and modelling

Abstract

In this paper, we extend the classical Lie symmetry analysis from partial differential equations to integro-differential equations with functional derivatives. We continue the work of OBERLACK and WACŁAWCZYK (2006, *Arch. Mech.*, 58, 597), (2013, *J. Math. Phys.*, 54, 072901) where the extended Lie symmetry analysis is performed in the Fourier space. Here, we introduce a method to perform the extended Lie symmetry analysis in the physical space where we have to deal with the transformation of the integration variable in the appearing integral terms. The method is based on the transformation of the product $y(x)\mathrm{d}x$ appearing in the integral terms and applied to the functional formulation of the viscous Burgers equation. The extended Lie symmetry analysis furnishes all known symmetries of the viscous Burgers equation and is able to provide new symmetries associated with the Hopf formulation of the viscous Burgers equation. Hence, it can be employed as an important tool for applications in continuum mechanics.

Contents

1 Introduction

1.1 Three complete descriptions of turbulence

Turbulence research knows three complete descriptions of turbulence. In each of these descriptions, the aim is to calculate the statistics of a turbulent flow by determining the so-called *multi-point velocity correlations* or shortly *multi-point correlations*

$$\left\langle U_{i_1}^t(\boldsymbol{x}_1)\ldots U_{i_n}^t(\boldsymbol{x}_n)\right\rangle, \quad i_1,\ldots,i_n = 1,2,3.$$

In this term, $\boldsymbol{U}^t$ is a random vector describing the velocity field in time t and $\boldsymbol{x}_1,\ldots,\boldsymbol{x}_n$ are positions of different fluid particles in the space occupied by the fluid. $\boldsymbol{U}^t$ is given by

$$\boldsymbol{U}^t(\boldsymbol{x}_i) = \left(U_1^t(\boldsymbol{x}_i), U_2^t(\boldsymbol{x}_i), U_3^t(\boldsymbol{x}_i)\right) = \left(U_1(\boldsymbol{x}_i, t), U_2(\boldsymbol{x}_i, t), U_3(\boldsymbol{x}_i, t)\right).$$

In the language of stochastics, the multi-point correlations are the covariances of the velocity components.

We briefly introduce the three complete descriptions of turbulence research.

- In the *multi-point correlation approach* an infinite dimensional chain of linear but non-local differential equations have to be solved. On the n-th level, the unknown $(n + 1)$-point correlation is present. Solving the infinitely many equations provides directly all multi-point correlations. In [3], the Lie symmetries of the infinite set of multi-point correlation equations are investigated.

- In the *Lundgren-Monin-Novikov approach* [5] it is assumed that the velocity field $\boldsymbol{U}^t$ admit probability density functions (PDF's)

$$f_n^t := f^t(\boldsymbol{v}_1,\ldots,\boldsymbol{v}_n;\boldsymbol{x}_1,\ldots,\boldsymbol{x}_n)$$

  given in terms of the Dirac delta distribution and describing the correlation of the velocity components at multiple points in space. To be more precise, $f^t(\boldsymbol{v}_1;\boldsymbol{x}_1)\,\mathrm{d}\boldsymbol{v}_1$ expresses the probability that the velocity vector $\boldsymbol{U}^t(\boldsymbol{x}_1) = \boldsymbol{U}(\boldsymbol{x}_1, t)$ is contained within the infinitesimal interval $[\boldsymbol{v}_1, \boldsymbol{v}_1 + \mathrm{d}\boldsymbol{v}_1]$. The *Lundgren-Monin-Novikov hierarchy* is an infinite dimensional chain of non-local differential equations for the PDF's where on the n-th level the unknown $(n + 1)$-point PDF $f_{n+1}^t = f^t(\boldsymbol{v}_1,\ldots,\boldsymbol{v}_{(n+1)};\boldsymbol{x}_1,\ldots,\boldsymbol{x}_{(n+1)})$ is present. Solving the infinitely many equations provides all PDF's. The multi-point correlations can be calculated by integrating the PDF's via

$$\left\langle U_{i_1}^t(\boldsymbol{x}_1)\ldots U_{i_n}^t(\boldsymbol{x}_n)\right\rangle = \int_{\mathbb{R}^3}\cdots\int_{\mathbb{R}^3} v_{i_1}\ldots v_{i_n} f^t(\boldsymbol{v}_1,\ldots,\boldsymbol{v}_n;\boldsymbol{x}_1,\ldots,\boldsymbol{x}_n)\,\mathrm{d}\boldsymbol{v}_1\ldots\mathrm{d}\boldsymbol{v}_n.$$

  In [4], the Lie symmetries of the Lundgren-Monin-Novikov hierarchy are investigated.

- In the *Hopf approach* the characteristic functions ϕ of the PDF's f^t for $n \to \infty$ are investigated, cf. Ref. [6]. The n-point characteristic function is defined as

$$\phi(\boldsymbol{y}_1,\ldots,\boldsymbol{y}_n, t) := \left\langle e^{i(\mathcal{U}^t, y)}\right\rangle = \int_{\mathbb{R}^3}\cdots\int_{\mathbb{R}^3} e^{i(v,y)} f^t(\boldsymbol{v}_1,\ldots,\boldsymbol{v}_n;\boldsymbol{x}_1,\ldots,\boldsymbol{x}_n)\,\mathrm{d}\boldsymbol{v}_1\ldots\mathrm{d}\boldsymbol{v}_n$$

with $\mathscr{U}^t := \big(U^t(x_1), \ldots, U^t(x_n)\big)$. $(\cdot, \cdot)$ is defined as the Euclidean scalar product

$$(v, y) := \sum_{k=1}^{n} \sum_{i=1}^{3} v_{k_i} y_{k_i}$$

for $v = (v_1, \ldots, v_n)$, $y = (y_1, \ldots, y_n)$ with $v_k, y_k \in \mathbb{R}^3$, $k = 1, \ldots, n$.

In hydromechanics, we assume that the mean free path is negligible, hence we take the continuum limit $n \to \infty$. Thus, instead of the n velocity vectors $v_1, \ldots, v_n$, we consider a continuous set $[v(x)]$ such that the velocity v is a continuous function depending on the spatial variable x. In the continuum limit, the probability density function f^t becomes a probability density functional and the function ϕ becomes the functional

$$\phi([y(x)], t) = \left\langle e^{i(U^t, y)} \right\rangle = \int e^{i(v, y)} f^t([v(x)]) \, d[v(x)]. \tag{1.1}$$

In this paper, we pursuit the Hopf approach defined by equation (1.1). A relation between the Hopf approach and the Lundgren-Monin-Novikov approach is discussed in Ref. [7] where also equations in the Lagrangian multi-particle framework are investigated.

1.2 Hopf functional and multi-point correlations

In equation (1.1) it is not directly apparent what is meant by $(\cdot, \cdot)$ and how the integration domain should be chosen. In order to define $\phi = \phi([y(x)], t)$ properly, we introduce the L^2 space.

Definition 1.2.1 (L^2 space). For $G \subseteq \mathbb{R}^3$ define

$$L^2(G, \mathbb{R}^3) := \left\{ v : G \longrightarrow \mathbb{R}^3 \,\middle|\, \int_G v(x) \cdot v(x) \, dx = \int_G \sum_{i=1}^{3} v_i(x) v_i(x) \, dx < \infty \right\}$$

and equip $L^2(G, \mathbb{R}^3)$ with $(\cdot, \cdot)$ defined by

$$(v, y) := \int_G v(x) \cdot y(x) \, dx = \int_G \sum_{i=1}^{3} v_i(x) y_i(x) \, dx \tag{1.2}$$

for $v, y \in L^2(G, \mathbb{R}^3)$.

If we equip $L^2(G, \mathbb{R}^3)$ with $(\cdot, \cdot)$ defined by equation (1.2), it is a Hilbert space, cf. Ref. [11].

Definition 1.2.2 (Hopf functional, Hopf functional differential equation). Let $y \in L^2(G, \mathbb{R}^3)$ and let $(\cdot, \cdot)$ be the L^2 scalar product defined by equation (1.2). The *characteristic functional* or *Hopf functional* is defined by

$$\phi([y(x)], t) := \left\langle e^{i(U^t, y)} \right\rangle = \int_{L^2(G, \mathbb{R}^3)} e^{i(v, y)} f^t([v(x)]) \, d[v(x)]. \tag{1.3}$$

Instead of dealing with an infinite dimensional chain of differential equations, the Hopf approach works with one scalar *functional differential equation* (FDE).

A FDE is a generalization of a partial differential equation (PDE). A PDE becomes a FDE if the number of independent variables tends to infinity. Precisely:

Definition 1.2.3 (Functional, functional derivative, functional differential equation).

- Let $\mathscr{F}$ be a function space. A *functional* is a mapping

$$\phi : \mathscr{F} \longrightarrow \mathbb{R}.$$

- Let $\phi = \phi([\mathbf{y}(\mathbf{x})])$ be a functional. We define[1] the *functional derivative* of ϕ as the limit

$$\frac{\partial \phi([\mathbf{y}(\mathbf{x})])}{\partial y_\alpha(\mathbf{z})\mathrm{d}\mathbf{z}} := \lim_{h \to 0} \frac{\phi([\mathbf{y}(\mathbf{x}) + h\mathbf{e}_\alpha \delta(\mathbf{x} - \mathbf{z})]) - \phi([\mathbf{y}(\mathbf{x})])}{h}, \qquad \alpha = 1,2,3.$$

Here, $\mathbf{e}_\alpha$ denotes the αth unit vector and δ denotes the Dirac delta distribution. We denote the functional derivative by

$$\phi_{,y_\alpha(\mathbf{z})} = \frac{\partial \phi}{\partial y_\alpha(\mathbf{z})\mathrm{d}\mathbf{z}} = \frac{\delta \phi}{\delta y_\alpha(\mathbf{z})}.$$

- Let $\phi = \phi([\mathbf{y}(\mathbf{x}')], \mathbf{x}, t)$ be a functional. A *functional differential equation* (FDE) of order q is an equation $F = 0$ where F is a functional relating ϕ and all its derivatives up to order q, which can include partial derivatives with respect to t, partial derivatives with respect to x and functional derivatives with respect to each $y_\alpha(\mathbf{x}')$, $\alpha = 1,2,3$:

$$F([\mathbf{y}(\mathbf{x}')], \mathbf{x}, t, \phi, \underset{1}{\phi}, \underset{2}{\phi}, \dots, \underset{q}{\phi}) = 0. \tag{1.4}$$

In a FDE, the finite set $(\mathbf{y}_1, \dots, \mathbf{y}_n) \in (\mathbb{R}^3)^n$ is replaced by an infinite set $[\mathbf{y}(\mathbf{x}')] \in \mathscr{F}$ with a continuous parameter $\mathbf{x}'$ representing the continuum analogon of the discrete counting parameter $k = 1, \dots, n$. Hence, the dependent variables are *functionals* as they depend on functions.

It is important to note that we may obtain all multi-point correlations by differentiating ϕ and evaluating the derivatives at $\mathbf{y} = \mathbf{0}$ via

$$\left\langle U_{\alpha_1}^t(\mathbf{x}_1) \dots U_{\alpha_n}^t(\mathbf{x}_n) \right\rangle = \frac{1}{i^n} \phi_{,y_{\alpha_n}(\mathbf{x}_n) \dots y_{\alpha_1}(\mathbf{x}_1)} \Big|_{\mathbf{y}=0}. \tag{1.5}$$

Consequently, the Hopf functional ϕ provides the full statistical description of the velocity field $\mathbf{U}^t$ since all multi-point correlations can be expressed in terms of ϕ and its functional derivatives.

In order to solve FDE's containing the Hopf functional ϕ, so-called *Hopf FDE's*, we use an extension of the classical Lie symmetry analysis. The extension is done in [1] where it is shown how the Lie symmetry analysis is performed upon FDE's. The extended Lie symmetry analysis is applied to the Hopf formulation of the viscous and inviscid Burgers equation in [2].

In order to circumvent the partial derivatives $\partial/\partial x$ appearing both in the viscous and in the inviscid Hopf-Burgers FDE, in [1] and [2] the extended Lie symmetry analysis is performed in the *Fourier* space by considering the Fourier transform of the Hopf-Burgers FDE. In the Fourier space, derivatives with respect to x become multipliers with ik. Hence, in Refs. [1] and [2] k was not transformed separately but was treated as a continuous index. In [2] we also derived invariant solutions for the Hopf equations, however, due to the restriction of the method (no transformation of k), the solutions contained unknown functions of k governed by an infinite chain of equations. We note that the Hopf functional was also considered in Refs. [8], [9] where, using numerical integration, scaling in k and decay of energy was calculated. As far as symmetry methods are concerned, interesting aspects of energy scaling for turbulence were discovered in [10] based on self-similar solutions of the Leith model.

---

[1] In fact, this is a corollary of the definition of the Gâteaux derivative, cf. Ref. [12]. As the presented formula is more suitable for calculations, we use it as definition.

In this paper, the extended Lie symmetry analysis is performed on the viscous Hopf-Burgers FDE in the *physical* space. Hence, we have to deal with partial derivatives with respect to x. There are several approaches how to do that. The approaches are introduced in section 3.1.

1.2.1 Viscous Hopf-Burgers functional integro-differential equation

Our primary long-term goal is to investigate the Hopf functional of turbulent velocity, i.e. the Hopf-Navier-Stokes FDE. For the sake of convenience, we presently restrict ourselves to the one-dimensional case, i.e. replace $\mathbb{R}^3$ by $\mathbb{R}$. Instead of the incompressible Navier-Stokes equations, we use the viscous Burgers equation to derive the so-called *viscous Hopf-Burgers FDE*. We assume that $y, U^t \in L^2(G, \mathbb{R})$ with

$$y = y(x), \qquad U^t = U^t(x) = U(x, t)$$

and that U^t fulfills the viscous Burgers equation

$$\frac{\partial U^t}{\partial t} + U^t \frac{\partial U^t}{\partial x} = \nu \frac{\partial^2 U^t}{\partial x^2}$$

which can equivalently by written as

$$\frac{\partial U^t}{\partial t} + \frac{1}{2} \frac{\partial (U^t)^2}{\partial x} = \nu \frac{\partial^2 U^t}{\partial x^2}.$$

By simple rescaling we may eliminate the factor $1/2$. We call the following equation viscous Burgers equation:

$$\frac{\partial U^t}{\partial t} + \frac{\partial (U^t)^2}{\partial x} = \nu \frac{\partial^2 U^t}{\partial x^2}. \tag{1.6}$$

We differentiate the Hopf functional (1.3) with respect to t to get

$$\phi_{,t}([y(x)], t) = \frac{\partial}{\partial t} \left\langle e^{i(U^t, y)} \right\rangle = \left\langle i \left(\frac{\partial U^t}{\partial t}, y \right) e^{i(U^t, y)} \right\rangle = \left\langle i \left(-\frac{\partial (U^t)^2}{\partial x} + \nu \frac{\partial^2 U^t}{\partial x^2} \right) e^{i(U^t, y)} \right\rangle \tag{1.7}$$

where equation (1.6) was inserted in order to eliminate $\partial U^t / \partial t$. Using the definition of the characteristic functional (1.1), the following relations can be calculated:

$$-\left\langle \frac{\partial (U^t)^2}{\partial x} e^{i(U^t, y)} \right\rangle = \frac{\partial}{\partial x} \frac{\delta^2 \phi}{\delta y(x) \delta y(x)} = \phi_{,xy(x)y(x)},$$

$$i\nu \left\langle \frac{\partial^2 U^t}{\partial x^2} e^{i(U^t, y)} \right\rangle = \nu \frac{\partial}{\partial x^2} \frac{\delta \phi}{\delta y(x)} = \nu \phi_{,xxy(x)}.$$

After introducing the above formulas into equation (1.7), we finally obtain the *viscous Hopf-Burgers FDE*

$$\boxed{\phi_{,t}([y(x)], t) = \int_G y(x) \left(i\phi_{,xy(x)y(x)} + \nu\phi_{,xxy(x)} \right) \mathrm{d}x} \tag{1.8}$$

with $y \in L^2(G, \mathbb{R})$. Since ϕ is part of the integrand, the viscous Hopf-Burgers FDE is an *integro-differential equation*.

In the following, we want to perform the extended Lie symmetry analysis on the viscous Hopf-Burgers FDE. Before doing that, we review the extended Lie symmetry analysis developed in [1] and [2] and advance it for our purposes.

This paper is structured as follows: In the second section, we present an extension of the classical Lie symmetry analysis based on the two papers [1] and [2] which allows us to analyze functional equations in the physical space. The third section presents the main part of the paper: We apply the extended Lie symmetry analysis to the viscous Hopf-Burgers functional integro-differential equation. First, we present three different approaches to perform the extended Lie symmetry analysis on the viscous Hopf-Burgers FDE. Subsequently, we solve the system of equations for the infinitesimals. Then, we discuss symmetry breaking restrictions and indicate physically relevant symmetries. In the end of the third section, we compare the symmetries of the viscous Hopf-Burgers FDE with the symmetries of the viscous Burgers equation and calculate the associated global transformations. Finally, a conclusion and perspectives are given in the fourth section.

2 Extension of the Lie symmetry analysis towards functional integro-differential equations

2.1 One-parameter Lie point transformations

To start with, we introduce some basic notions, cf. Ref. [1]. As in this paper we consider the viscous Hopf-Burgers FDE being of third order, we set $q = 3$ in equation (1.4). Furthermore, as we use the viscous Burgers equation as the underlying equation for fluid motion instead of the incompressible Navier-Stokes equations, we replace $\mathbb{R}^3$ by $\mathbb{R}$. The whole theory can easily be extended to higher dimensions and to differential equations of arbitrary order q.

In Lie symmetry analysis, we only consider *continuous* symmetry transformations depending on a continuous parameter $\varepsilon \in S$ where S is a Lie group. We restrict ourselves to so-called *one-paramater Lie point transformations*, cf. Ref. [14]:

Definition 2.1.1 (One-parameter Lie point transformations of PDE's). Consider the PDE

$$F(y_1, \ldots, y_n, x_1, \ldots, x_n, t, \phi, \underset{1}{\phi}, \underset{2}{\phi}, \ldots, \underset{3}{\phi}) = 0$$

and let

$$
\begin{aligned}
\overline{y}_i &= \overline{y}_i(y_1, \ldots, y_n, x_1, \ldots, x_n, t, \phi, \varepsilon), & i &= 1, \ldots, n, \\
\overline{x}_i &= \overline{x}_i(y_1, \ldots, y_n, x_1, \ldots, x_n, t, \phi, \varepsilon), & i &= 1, \ldots, n, \\
\overline{t} &= \overline{t}(y_1, \ldots, y_n, x_1, \ldots, x_n, t, \phi, \varepsilon), & & \\
\overline{\phi} &= \overline{\phi}(y_1, \ldots, y_n, x_1, \ldots, x_n, t, \phi, \varepsilon) & &
\end{aligned}
$$

be the transformed variables. The transformation is called *one-parameter Lie point transformation* if and only if the transformed variables are given by

$$
\begin{aligned}
\overline{y}_i &= y_i + \xi_{y_i}(y_1, \ldots, y_n, x_1, \ldots, x_n, t, \phi)\varepsilon, & i &= 1, \ldots, n, \\
\overline{x}_i &= x_i + \xi_{x_i}(y_1, \ldots, y_n, x_1, \ldots, x_n, t, \psi)\varepsilon, & i &= 1, \ldots, n, \\
\overline{t} &= t + \xi_t(y_1, \ldots, y_n, x_1, \ldots, x_n, t, \phi)\varepsilon, & & \\
\overline{\phi} &= \phi + \eta_\phi(y_1, \ldots, y_n, x_1, \ldots, x_n, t, \phi)\varepsilon. & &
\end{aligned}
$$

In a one-parameter Lie point transformation, the derivatives of ϕ are not transformed separately. The transformations of $\underset{1}{\phi}, \underset{2}{\phi}, \ldots, \underset{q}{\phi}$ are calculated as functions of $\xi_{y_1}, \ldots, \xi_{y_n}, \xi_{x_1}, \ldots, \xi_{x_n}, \xi_t, \eta_\phi$. We write

$$\overline{\phi}_{,\overline{t}} = \phi_{,t} + \underset{1}{\zeta_{;t}}(y_1, \ldots, y_n, x_1, \ldots, x_n, t, \phi, \phi)\varepsilon,$$

$$\overline{\phi}_{,\overline{y}_i} = \phi_{,y_i} + \underset{1}{\zeta_{;y_i}}(y_1, \ldots, y_n, x_1, \ldots, x_n, t, \phi, \phi)\varepsilon, \qquad i = 1, \ldots, n,$$

$$\overline{\phi}_{,\overline{y}_i \overline{y}_j} = \phi_{,y_i y_j} + \underset{1}{\zeta_{;y_i y_j}}(y_1, \ldots, y_n, x_1, \ldots, x_n, t, \phi, \underset{1}{\phi}, \underset{2}{\phi})\varepsilon, \qquad i, j = 1, \ldots, n,$$

$$\vdots$$

The functions $\xi_{y_i}, \xi_{x_i}, \xi_t, \eta_\phi, \zeta_{;\ldots}$ are called *infinitesimals*. Notice that we separate indices of ζ by a semicolon to distinguish them from derivatives. The transformations are expanded in a Taylor series

$$(T\overline{f})_{\varepsilon_0}(\varepsilon) := \sum_{k=0}^{\infty} \frac{1}{k!} \left.\frac{\partial^k \overline{f}}{\partial \varepsilon^k}\right|_{\varepsilon=\varepsilon_0} (\varepsilon - \varepsilon_0)^k \tag{2.1}$$

about $\varepsilon_0 = 0$, hence the infinitesimals are defined by

$$\xi_{y_i} := \left.\frac{\partial \overline{y_i}}{\partial \varepsilon}\right|_{\varepsilon=0}, \quad \xi_{x_i} := \left.\frac{\partial \overline{x_i}}{\partial \varepsilon}\right|_{\varepsilon=0}, \quad \xi_t := \left.\frac{\partial \overline{t}}{\partial \varepsilon}\right|_{\varepsilon=0}, \quad \eta_\phi := \left.\frac{\partial \overline{\phi}}{\partial \varepsilon}\right|_{\varepsilon=0},$$

$$\zeta_{;t} := \left.\frac{\partial \overline{\phi}_{,\overline{t}}}{\partial \varepsilon}\right|_{\varepsilon=0}, \quad \zeta_{;y_i} := \left.\frac{\partial \overline{\phi}_{,\overline{y_i}}}{\partial \varepsilon}\right|_{\varepsilon=0}, \quad \ldots.$$

We extend this definition to FDE's.

Definition 2.1.2 (One-parameter Lie point transformations of FDE's). Consider the FDE

$$F([y(x')], x, t, \phi, \underset{1}{\phi}, \underset{2}{\phi}, \underset{3}{\phi}) = 0$$

and let

$$\begin{aligned}
\overline{y(z)\mathrm{d}z} &= \overline{y(z)\mathrm{d}z}([y(x')], z, x, t, \phi, \varepsilon), & x', z \in G, \\
\overline{x} &= \overline{x}([y(x')], x, t, \phi, \varepsilon), & x' \in G, \\
\overline{t} &= \overline{t}([y(x')], x, t, \phi, \varepsilon), & x' \in G, \\
\overline{\phi} &= \overline{\phi}([y(x')], x, t, \phi, \varepsilon), & x' \in G,
\end{aligned}$$

be the transformed variables where $G \subseteq \mathbb{R}$ is an uncountable set. The transformation is called *one-parameter Lie point transformation* if and only if the transformed variables are given by

$$\begin{aligned}
\overline{y(z)\mathrm{d}z} &= y(z)\mathrm{d}z + \xi_{y(z)}\mathrm{d}z([y(x')], z, x, t, \phi)\varepsilon, & x', z \in G, \\
\overline{x} &= x + \xi_x([y(x')], x, t, \phi)\varepsilon, & x' \in G, \\
\overline{t} &= t + \xi_t([y(x')], x, t, \phi)\varepsilon, & x' \in G, \\
\overline{\phi} &= \phi + \eta_\phi([y(x')], x, t, \phi)\varepsilon, & x' \in G.
\end{aligned}$$

Again, the derivatives of ϕ are not transformed separately. The transformations of $\underset{1}{\phi}, \underset{2}{\phi}, \underset{3}{\phi}$ are calculated as functionals depending on $\xi_{y(z)}\mathrm{d}z, \xi_x, \xi_t, \eta_\phi$. We write

$$\overline{\phi}_{,\overline{t}} = \phi_{,t} + \zeta_{;t}([y(x')], x, t, \phi, \underset{1}{\phi})\varepsilon, \qquad\qquad x' \in G,$$

$$\overline{\phi}_{,\overline{y(z_1)}} = \phi_{,y(z_1)} + \zeta_{;y(z_1)}([y(x')], z_1, x, t, \phi, \underset{1}{\phi})\varepsilon, \qquad x', z_1 \in G,$$

$$\overline{\phi}_{,\overline{y(z_2)}\,\overline{y(z_1)}} = \phi_{,y(z_2)y(z_1)} + \zeta_{;y(z_2)y(z_1)}([y(x')], z_1, z_2, x, t, \phi, \underset{1}{\phi}, \underset{2}{\phi})\varepsilon, \qquad x', z_1, z_2 \in G,$$

$$\vdots$$

The functionals $\xi_{y(z)}\mathrm{d}z, \xi_x, \xi_t, \eta_\phi, \zeta_{;t}, \ldots$ are called *infinitesimals*. Again, we separate indices of ζ by a semicolon to distinguish them from derivatives. The transformations are expanded in a Taylor series (2.1) about $\varepsilon_0 = 0$, hence the infinitesimals are defined by

$$\xi_{y(z)}\mathrm{d}z := \left.\frac{\partial \overline{y(z)\mathrm{d}z}}{\partial \varepsilon}\right|_{\varepsilon=0}, \quad \xi_x := \left.\frac{\partial \overline{x}}{\partial \varepsilon}\right|_{\varepsilon=0}, \quad \xi_t := \left.\frac{\partial \overline{t}}{\partial \varepsilon}\right|_{\varepsilon=0}, \quad \eta_\phi := \left.\frac{\partial \overline{\phi}}{\partial \varepsilon}\right|_{\varepsilon=0},$$

$$\zeta_{;t} := \left.\frac{\partial \overline{\phi}_{,\overline{t}}}{\partial \varepsilon}\right|_{\varepsilon=0}, \quad \zeta_{;y(z)} := \left.\frac{\partial \overline{\phi}_{,\overline{y(z)}}}{\partial \varepsilon}\right|_{\varepsilon=0}, \quad \ldots.$$

Comparing the notion of one-parameter Lie point transformations of PDE's and of FDE's, one sees the considered transformed functions $\overline{y_i}$, $\overline{x_i}$, $\overline{t}$, $\overline{\phi}$ are replaced by functionals $\overline{y(z)\mathrm{d}z}$, $\overline{x}$, $\overline{t}$, $\overline{\phi}$ depending on the infinite set $[y(x')]$. Notice that we do not transform $y(z)$ but $y(z)\mathrm{d}z$ as

$$\int_G y(z)\,\mathrm{d}z \text{ is the continuum analogon of } \sum_{i=1}^{n} y_i.$$

Following Ref. [16], the variable $y(z)\mathrm{d}z$ can be represented as a test series

$$y(z)\mathrm{d}z = \sum_{n=1}^{\infty} y_n h_n(z)\mathrm{d}z. \tag{2.2}$$

where $\{h_n(z)\}_{n\in\mathbb{N}}$ is a set of orthogonal functions. In this respect, the transformed variable $\overline{y(z)\mathrm{d}z}$ reads

$$\overline{y(z)\mathrm{d}z} = \sum_{n=1}^{\infty} \overline{y_n} h_n(\overline{z})\mathrm{d}\overline{z}. \tag{2.3}$$

A possible approach in extending the Lie group analysis would be to account for the transformations of y_n and z separately. However, due to the presence of functional derivatives in equation (1.8), i.e. derivatives with respect to $y(z)\mathrm{d}z$, we prefer to consider $y(z)\mathrm{d}z$ as a variable to be transformed. Still, the decomposition (2.2) will be taken into account in the definitions of the infinitesimals in section 2.3.

An additional option would be to transform $y(z)$ instead of $y(z)\mathrm{d}z$. Then, one has to take into consideration the transformation of $\mathrm{d}z$. There are two ways how this could be done, cf. section 3.1; however, we do not pursuit those methods here. In order to be consistent, the infinitesimal associated with the transformation $\overline{y(z)\mathrm{d}z}$ is called $\xi_{y(z)\mathrm{d}z}$ instead of $\xi_{y(z)}$. The infinitesimal $\xi_{y(z)\mathrm{d}z}$ has to depend explicitly on z as it defines a new variable for each $z \in G$. Analogous considerations hold true for the transformations of the functional derivatives of the dependent variable $\overline{\phi}_{,\ldots}$ and their infinitesimals $\zeta_{;\ldots}$.

2.2 Differential operators

In the classical Lie symmetry analysis, the variables $(y_1,\ldots,y_n, x_1,\ldots,x_n, t, \phi, \underset{1}{\phi}, \underset{2}{\phi}, \underset{3}{\phi})$ are treated as independent variables. Hence, in the extended Lie symmetry analysis, we treat the variables $([y(x')], x, t, \phi, \underset{1}{\phi}, \underset{2}{\phi}, \underset{3}{\phi})$ as independent variables. The Hopf functional ϕ does not depend explicitly on x, hence all derivatives of ϕ with respect to x vanish except when ϕ is derived first with respect to $y(x')$ and then with respect to x. Neglecting the vanishing summands and considering the non-commutativity

$$\frac{\partial^2}{\partial x\, \partial y(x)\mathrm{d}x} \neq \frac{\partial^2}{\partial y(x)\mathrm{d}x\, \partial x},$$

we can define:

Definition 2.2.1 (Differential operators for Hopf FDE's). For a Hopf FDE of third order

$$F([y(x')], x, t, \phi, \underset{1}{\phi}, \underset{2}{\phi}, \underset{3}{\phi}) = 0$$

we introduce

$$\frac{\mathscr{D}}{\mathscr{D}t} = \frac{\partial}{\partial t} + \phi_{,t}\frac{\partial}{\partial\phi} + \phi_{,tt}\frac{\partial}{\partial\phi_{,t}} + \int_G \phi_{,ty(x')}\mathrm{d}x'\frac{\partial}{\partial\phi_{,y(x')}\mathrm{d}x'} + \phi_{,ttt}\frac{\partial}{\partial\phi_{,tt}}$$
$$+ \int_G \phi_{,tty(x')}\mathrm{d}x'\frac{\partial}{\partial\phi_{,ty(x')}\mathrm{d}x'} + \int_G \phi_{,txy(x')}\mathrm{d}x'\frac{\partial}{\partial\phi_{,xy(x')}\mathrm{d}x'}$$
$$+ \int_G\int_G \phi_{,ty(x'')y(x')}\mathrm{d}x'\mathrm{d}x''\frac{\partial}{\partial\phi_{,y(x'')y(x')}\mathrm{d}x'\mathrm{d}x''},$$

$$\frac{\mathscr{D}}{\mathscr{D}x} = \frac{\partial}{\partial x} + \int_G \phi_{,xy(x')}\mathrm{d}x'\frac{\partial}{\partial\phi_{,y(x')}\mathrm{d}x'} + \int_G \phi_{,xty(x')}\mathrm{d}x'\frac{\partial}{\partial\phi_{,ty(x')}\mathrm{d}x'}$$
$$+ \int_G \phi_{,xxy(x')}\mathrm{d}x'\frac{\partial}{\partial\phi_{,xy(x')}\mathrm{d}x'} + \int_G\int_G \phi_{,xy(x'')y(x')}\mathrm{d}x'\mathrm{d}x''\frac{\partial}{\partial\phi_{,y(x'')y(x')}\mathrm{d}x'\mathrm{d}x''},$$

$$\frac{\mathscr{D}}{\mathscr{D}y(z)\mathrm{d}z} = \frac{\partial}{\partial y(z)\mathrm{d}z} + \phi_{,y(z)}\frac{\partial}{\partial\phi} + \phi_{,y(z)t}\frac{\partial}{\partial\phi_{,t}} + \int_G \phi_{,y(z)y(x')}\mathrm{d}x'\frac{\partial}{\partial\phi_{,y(x')}\mathrm{d}x'} + \phi_{,y(z)tt}\frac{\partial}{\partial\phi_{,tt}}$$
$$+ \int_G \phi_{,y(z)ty(x')}\mathrm{d}x'\frac{\partial}{\partial\phi_{,ty(x')}\mathrm{d}x'} + \int_G \phi_{,y(z)xy(x')}\mathrm{d}x'\frac{\partial}{\partial\phi_{,xy(x')}\mathrm{d}x'}$$
$$+ \int_G\int_G \phi_{,y(z)y(x'')y(x')}\mathrm{d}x'\mathrm{d}x''\frac{\partial}{\partial\phi_{,y(x'')y(x')}\mathrm{d}x'\mathrm{d}x''}.$$

2.3 Infinitesimals

By means of the differential operators presented in definition 2.2.1, we can calculate the infinitesimals $\zeta_{;\dots}$ as functionals of the infinitesimals $\xi_{y(z)}\mathrm{d}z, \xi_x, \xi_t, \eta_\phi$. For the viscous Hopf-Burgers FDE, we need the three infinitesimals $\zeta_{;t}$, $\zeta_{;xy(x)y(x)}$ and $\zeta_{;xxy(x)}$.

- In order to calculate $\zeta_{;t}$, we differentiate the transformed Hopf functional $\overline{\phi}$ with respect to t taking into account the decomposition (2.2):

$$\frac{\mathscr{D}\overline{\phi}}{\mathscr{D}t} = \frac{\mathscr{D}\overline{\phi}}{\mathscr{D}\overline{t}}\frac{\mathscr{D}\overline{t}}{\mathscr{D}t} + \frac{\mathscr{D}\overline{\phi}}{\mathscr{D}\overline{x}}\frac{\mathscr{D}\overline{x}}{\mathscr{D}t} + \int_{\overline{G}} \frac{\mathscr{D}\overline{\phi}}{\mathscr{D}\overline{y}(x')\mathrm{d}x'}\sum_{n=1}^{\infty}\frac{\mathscr{D}\overline{y}_n}{\mathscr{D}t}h_n(\overline{x'})\mathrm{d}\overline{x'}. \tag{2.4}$$

Hence, in the definition above we account for the fact that t, x and the infinite set $\{y_n\}_{n\in\mathbb{N}}$ are the independent variables. Still, as argued in section 2.1, in the extended Lie group analysis we transform the continuum variable $y(x)\mathrm{d}x$, hence we will express the last RHS term in equation (2.4) in terms of $y(x)\mathrm{d}x$. This step is crucial for the further successful recovery of symmetries of the Burgers equation. Using (2.2) again, the time differential operator applied to $\overline{y(x')\mathrm{d}x'}$ can be decomposed into

$$\frac{\mathscr{D}\overline{y(x)\mathrm{d}x}}{\mathscr{D}t} = \sum_{n=1}^{\infty}\frac{\mathscr{D}\overline{y}_n}{\mathscr{D}t}h_n(\overline{x})\mathrm{d}\overline{x} + \sum_{n=1}^{\infty}y_n\frac{\mathscr{D}h_n(\overline{x})\mathrm{d}\overline{x}}{\mathscr{D}t}. \tag{2.5}$$

The second RHS term can be rewritten as

$$\sum_{n=1}^{\infty}\frac{\mathscr{D}\overline{y}_n}{\mathscr{D}t}h_n(\overline{x})\mathrm{d}\overline{x} = \frac{\mathscr{D}\overline{y(x)\mathrm{d}x}}{\mathscr{D}t} - \frac{\mathscr{D}\overline{y(x)\mathrm{d}x}}{\mathscr{D}\overline{x}}\frac{\mathscr{D}\overline{x}}{\mathscr{D}t} \tag{2.6}$$

and substituted into equation (2.4). Taking into account the fact that $\overline{\phi}$ does not depend explicitely on $\overline{x}$, from equation (2.4) we obtain

$$\frac{\mathscr{D}\overline{\phi}}{\mathscr{D}t} = \frac{\mathscr{D}\overline{\phi}}{\mathscr{D}\overline{t}}\frac{\mathscr{D}\overline{t}}{\mathscr{D}t} + \int_{\overline{G}}\frac{\mathscr{D}\overline{\phi}}{\mathscr{D}\overline{y(x')}dx'}\frac{\mathscr{D}\overline{y(x')}dx'}{\mathscr{D}t} - \int_{\overline{G}}\frac{\mathscr{D}\overline{\phi}}{\mathscr{D}\overline{y(x')}dx'}\frac{\mathscr{D}\overline{x'}}{\mathscr{D}t}\frac{\mathscr{D}\overline{y(x')}dx'}{\mathscr{D}\overline{x'}}. \qquad (2.7)$$

We integrate the last term by parts and assume that ϕ and its derivatives are zero at the boundaries. This step is necessary, as we want to introduce Lie point transformations, cf. definition 2.1.2, for the quantities in the integral. We finally obtain

$$\frac{\mathscr{D}\overline{\phi}}{\mathscr{D}t} = \frac{\mathscr{D}\overline{\phi}}{\mathscr{D}\overline{t}}\frac{\mathscr{D}\overline{t}}{\mathscr{D}t} + \int_{\overline{G}}\frac{\mathscr{D}\overline{\phi}}{\mathscr{D}\overline{y(x')}dx'}\frac{\mathscr{D}\overline{y(x')}dx'}{\mathscr{D}t} + \int_{\overline{G}}\frac{\mathscr{D}}{\mathscr{D}\overline{x'}}\left(\frac{\mathscr{D}\overline{\phi}}{\mathscr{D}\overline{y(x')}dx'}\right)\frac{\mathscr{D}\overline{x'}}{\mathscr{D}t}\overline{y(x')}dx'. \qquad (2.8)$$

With the one-parameter Lie point transformations, cf. definition 2.1.2, equation (2.8) reads

$$\begin{aligned}
\frac{\mathscr{D}(\phi + \varepsilon\eta_\phi)}{\mathscr{D}t} &= (\phi_{,t} + \varepsilon\zeta_{;t})\frac{\mathscr{D}(t + \varepsilon\xi_t)}{\mathscr{D}t} + \int_G (\phi_{,y(x')} + \varepsilon\zeta_{;y(x')})\frac{\mathscr{D}(y(x')dx' + \varepsilon\xi_{y(x')}dx')}{\mathscr{D}t} \\
&\quad + \int_G (\phi_{,x'y(x')} + \varepsilon\zeta_{;x'y(x')})\frac{\mathscr{D}(x' + \varepsilon\xi_{x'})}{\mathscr{D}t}\left(y(x')dx' + \varepsilon\xi_{y(x')}dx'\right). \qquad (2.9)
\end{aligned}$$

Evaluating this equation in $\mathcal{O}(1)$, we get

$$\frac{\mathscr{D}\phi}{\mathscr{D}t} = \phi_{,t}.$$

Using this, evaluating in $\mathcal{O}(\varepsilon)$ leads to

$$\frac{\mathscr{D}\eta_\phi}{\mathscr{D}t} = \phi_{,t}\frac{\mathscr{D}\xi_t}{\mathscr{D}t} + \zeta_{;t} + \int_G \phi_{,y(x')}\frac{\mathscr{D}\xi_{y(x')}dx'}{\mathscr{D}t} + \int_G \phi_{,x'y(x')}\frac{\mathscr{D}\xi_{x'}}{\mathscr{D}t}y(x')dx'$$

which constitutes an equation for $\zeta_{;t}$:

$$\zeta_{;t} = \frac{\mathscr{D}\eta_\phi}{\mathscr{D}t} - \phi_{,t}\frac{\mathscr{D}\xi_t}{\mathscr{D}t} - \int_G \phi_{,y(x')}\frac{\mathscr{D}\xi_{y(x')}dx'}{\mathscr{D}t} - \int_G \phi_{,x'y(x')}\frac{\mathscr{D}\xi_{x'}}{\mathscr{D}t}y(x')dx'.$$

Since the appearing infinitesimals $\xi_{y(x')}dx'$, ξ_t, η_ϕ do not depend on derivatives of ϕ, we immediately get

$$\begin{aligned}
\zeta_{;t} &= \frac{\partial\eta_\phi}{\partial t} + \phi_{,t}\left(\frac{\partial\eta_\phi}{\partial\phi} - \frac{\partial\xi_t}{\partial t}\right) - (\phi_{,t})^2\frac{\partial\xi_t}{\partial\phi} - \int_G \phi_{,y(x')}\frac{\partial\xi_{y(x')}dx'}{\partial t} - \int_G \phi_{,y(x')}\phi_{,t}\frac{\partial\xi_{y(x')}dx'}{\partial\phi} \\
&\quad - \int_G \phi_{,x'y(x')}\frac{\partial\xi_{x'}}{\partial t}y(x')dx' - \int_G \phi_{,x'y(x')}\phi_{,t}\frac{\partial\xi_{x'}}{\partial\phi}y(x')dx'. \qquad (2.10)
\end{aligned}$$

- In order to calculate $\zeta_{;y(x)}$, we differentiate $\overline{\phi}$ with respect to $y(x)$. An analog calculation leads to

$$\begin{aligned}
\zeta_{;y(x)} &= \frac{\mathscr{D}\eta_\phi}{\mathscr{D}y(x)dx} - \phi_{,t}\frac{\mathscr{D}\xi_t}{\mathscr{D}y(x)dx} - \int_G \phi_{,y(x')}\frac{\mathscr{D}\xi_{y(x')}dx'}{\mathscr{D}y(x)dx} - \int \phi_{,x'y(x')}\frac{\mathscr{D}\xi_{x'}}{\mathscr{D}y(x)dx}y(x')dx' \\
&= \frac{\partial\eta_\phi}{\partial y(x)dx} + \phi_{,y(x)}\frac{\partial\eta_\phi}{\partial\phi} - \phi_{,t}\frac{\partial\xi_t}{\partial y(x)dx} - \phi_{,t}\phi_{,y(x)}\frac{\partial\xi_t}{\partial\phi} - \int_G \phi_{,y(x')}\frac{\partial\xi_{y(x')}dx'}{\partial y(x)dx} \\
&\quad - \int_G \phi_{,y(x')}\phi_{,y(x)}\frac{\partial\xi_{y(x')}dx'}{\partial\phi} - \int \phi_{,x'y(x')}\frac{\partial\xi_{x'}}{\partial y(x)dx}y(x')dx' - \int \phi_{,x'y(x')}\phi_{,y(x)}\frac{d\xi_{x'}}{\partial\phi}y(x')dx'.
\end{aligned}$$

In order to calculate $\zeta_{;y(x)y(x)}$, we differentiate $\overline{\phi}_{,\overline{y(x)}}$ with respect to $y(x)$. As $\overline{\phi}_{,\overline{y(x)}}$ *does* depend on $\overline{x}$, we have

$$\frac{\mathscr{D}\overline{\phi}_{,\overline{y(x)}}}{\mathscr{D}y(x)\mathrm{d}x} = \frac{\mathscr{D}\overline{\phi}_{,\overline{y(x)}}}{\mathscr{D}\overline{t}}\frac{\mathscr{D}\overline{t}}{\mathscr{D}y(x)\mathrm{d}x} + \int_{\overline{G}}\frac{\mathscr{D}\overline{\phi}_{,\overline{y(x)}}}{\mathscr{D}\overline{y(x')}\mathrm{d}x'}\left(\frac{\mathscr{D}\overline{y(x')}\mathrm{d}x'}{\mathscr{D}y(x)\mathrm{d}x} - \frac{\mathscr{D}\overline{y(x')}\mathrm{d}x'}{\mathscr{D}\overline{x'}}\frac{\mathscr{D}\overline{x'}}{\mathscr{D}y(x)\mathrm{d}x}\right)$$

$$+ \frac{\mathscr{D}\overline{\phi}_{,\overline{y(x)}}}{\mathscr{D}\overline{x}}\frac{\mathscr{D}\overline{x}}{\mathscr{D}y(x)\mathrm{d}x}$$

$$= \overline{\phi}_{,\overline{t}\,\overline{y(x)}}\frac{\mathscr{D}\overline{t}}{\mathscr{D}y(x)\mathrm{d}x} + \int_{\overline{G}}\overline{\phi}_{,\overline{y(x')}\,\overline{y(x)}}\frac{\mathscr{D}\overline{y(x')}\mathrm{d}x'}{\mathscr{D}y(x)\mathrm{d}x} + \int_{G}\overline{\phi}_{,\overline{x'}\,\overline{y(x')}\,\overline{y(x)}}\frac{\mathscr{D}\overline{x'}}{\mathscr{D}y(x)\mathrm{d}x}\overline{y(x')}\mathrm{d}x'$$

$$+ \overline{\phi}_{,\overline{x}\,\overline{y(x)}}\frac{\mathscr{D}\overline{x}}{\mathscr{D}y(x)\mathrm{d}x}.$$

If we insert the one-parameter Lie point transformations, cf. definition 2.1.2, we get

$$\zeta_{;y(x)y(x)} = \frac{\mathscr{D}\zeta_{;y(x)}}{\mathscr{D}y(x)\mathrm{d}x} - \phi_{,ty(x)}\frac{\mathscr{D}\xi_t}{\mathscr{D}y(x)\mathrm{d}x} - \int_{G}\phi_{,y(x')y(x)}\frac{\mathscr{D}\xi_{y(x')}\mathrm{d}x'}{\mathscr{D}y(x)\mathrm{d}x} - \phi_{,xy(x)}\frac{\mathscr{D}\xi_x}{\mathscr{D}y(x)\mathrm{d}x}$$

$$- \int_{G}\phi_{,x'y(x')y(x)}\frac{\mathscr{D}\xi_{x'}}{\mathscr{D}y(x)\mathrm{d}x}y(x')\mathrm{d}x'.$$

In order to get $\zeta_{;xy(x)y(x)}$, we differentiate $\overline{\phi}_{,\overline{y(x)}\,\overline{y(x)}}$ with respect to x:

$$\frac{\mathscr{D}\overline{\phi}_{,\overline{y(x)}\,\overline{y(x)}}}{\mathscr{D}x} = \frac{\mathscr{D}\overline{\phi}_{,\overline{y(x)}\,\overline{y(x)}}}{\mathscr{D}\overline{t}}\frac{\mathscr{D}\overline{t}}{\mathscr{D}x} + \int_{\overline{G}}\frac{\mathscr{D}\overline{\phi}_{,\overline{y(x)}\,\overline{y(x)}}}{\mathscr{D}\overline{y(x')}\mathrm{d}x'}\frac{\mathscr{D}\overline{y(x')}\mathrm{d}x'}{\mathscr{D}x} + \frac{\mathscr{D}\overline{\phi}_{,\overline{y(x)}\,\overline{y(x)}}}{\mathscr{D}\overline{x}}\frac{\mathscr{D}\overline{x}}{\mathscr{D}x}$$

$$= \overline{\phi}_{,\overline{t}\,\overline{y(x)}\,\overline{y(x)}}\frac{\mathscr{D}\overline{t}}{\mathscr{D}x} + \int_{\overline{G}}\overline{\phi}_{,\overline{y(x')}\,\overline{y(x)}\,\overline{y(x)}}\frac{\mathscr{D}\overline{y(x')}\mathrm{d}x'}{\mathscr{D}x} + \overline{\phi}_{,\overline{x}\,\overline{y(x)}\,\overline{y(x)}}\frac{\mathscr{D}\overline{x}}{\mathscr{D}x}.$$

If we insert the one-parameter Lie point transformations, cf. definition 2.1.2, we get

$$\zeta_{;xy(x)y(x)} = \frac{\mathscr{D}\zeta_{;y(x)y(x)}}{\mathscr{D}x} - \phi_{,ty(x)y(x)}\frac{\mathscr{D}\xi_t}{\mathscr{D}x} - \int_{G}\phi_{,y(x')y(x)y(x)}\frac{\mathscr{D}\xi_{y(x')}\mathrm{d}x'}{\mathscr{D}x} - \phi_{,xy(x)y(x)}\frac{\mathscr{D}\xi_x}{\mathscr{D}x}. \quad (2.11)$$

- In order to calculate $\zeta_{;xy(x)}$, we differentiate $\overline{\phi}_{,\overline{y(x)}}$ with respect to x. We have

$$\frac{\mathscr{D}\overline{\phi}_{,\overline{y(x)}}}{\mathscr{D}x} = \frac{\mathscr{D}\overline{\phi}_{,\overline{y(x)}}}{\mathscr{D}\overline{t}}\frac{\mathscr{D}\overline{t}}{\mathscr{D}x} + \int_{\overline{G}}\frac{\mathscr{D}\overline{\phi}_{,\overline{y(x)}}}{\mathscr{D}\overline{y(x')}\mathrm{d}x'}\frac{\mathscr{D}\overline{y(x')}\mathrm{d}x'}{\mathscr{D}x} + \frac{\mathscr{D}\overline{\phi}_{,\overline{y(x)}}}{\mathscr{D}\overline{x}}\frac{\mathscr{D}\overline{x}}{\mathscr{D}x}$$

$$= \overline{\phi}_{,\overline{t}\,\overline{y(x)}}\frac{\mathscr{D}\overline{t}}{\mathscr{D}x} + \int_{\overline{G}}\overline{\phi}_{,\overline{y(x')}\,\overline{y(x)}}\frac{\mathscr{D}\overline{y(x')}\mathrm{d}x'}{\mathscr{D}x} + \overline{\phi}_{,\overline{x}\,\overline{y(x)}}\frac{\mathscr{D}\overline{x}}{\mathscr{D}x}.$$

If we insert the one-parameter Lie point transformations, cf. definition 2.1.2, we get

$$\zeta_{;xy(x)} = \frac{\mathscr{D}\zeta_{;y(x)}}{\mathscr{D}x} - \phi_{,ty(x)}\frac{\mathscr{D}\xi_t}{\mathscr{D}x} - \int_{G}\phi_{,y(x')y(x)}\frac{\mathscr{D}\xi_{y(x')}\mathrm{d}x'}{\mathscr{D}x} - \phi_{,xy(x)}\frac{\mathscr{D}\xi_x}{\mathscr{D}x}.$$

In order to get $\zeta_{;xxy(x)}$, we differentiate $\overline{\phi}_{,\overline{x}\,\overline{y(x)}}$ with respect to x:

$$\frac{\mathscr{D}\overline{\phi}_{,\overline{x}\,\overline{y(x)}}}{\mathscr{D}x} = \frac{\mathscr{D}\overline{\phi}_{,\overline{x}\,\overline{y(x)}}}{\mathscr{D}\overline{t}}\frac{\mathscr{D}\overline{t}}{\mathscr{D}x} + \int_{\overline{G}}\frac{\mathscr{D}\overline{\phi}_{,\overline{x}\,\overline{y(x)}}}{\mathscr{D}\overline{y(x')}\mathrm{d}x'}\frac{\mathscr{D}\overline{y(x')}\mathrm{d}x'}{\mathscr{D}x} + \frac{\mathscr{D}\overline{\phi}_{,\overline{x}\,\overline{y(x)}}}{\mathscr{D}\overline{x}}\frac{\mathscr{D}\overline{x}}{\mathscr{D}x}$$

$$= \overline{\phi}_{,\overline{t}\,\overline{x}\,\overline{y(x)}}\frac{\mathscr{D}\overline{t}}{\mathscr{D}x} + \int_{\overline{G}}\overline{\phi}_{,\overline{y(x')}\,\overline{x}\,\overline{y(x)}}\frac{\mathscr{D}\overline{y(x')}\mathrm{d}x'}{\mathscr{D}x} + \overline{\phi}_{,\overline{x}\,\overline{x}\,\overline{y(x)}}\frac{\mathscr{D}\overline{x}}{\mathscr{D}x}.$$

If we insert the one-parameter Lie point transformations, cf. definition 2.1.2, we get

$$\zeta_{;xxy(x)} = \frac{\mathscr{D}\zeta_{;xy(x)}}{\mathscr{D}x} - \phi_{,txy(x)}\frac{\mathscr{D}\xi_t}{\mathscr{D}x} - \int_G \phi_{,y(x')xy(x)}\frac{\mathscr{D}\xi_{y(x')}\mathrm{d}x'}{\mathscr{D}x} - \phi_{,xxy(x)}\frac{\mathscr{D}\xi_x}{\mathscr{D}x}. \tag{2.12}$$

Applying the differential operators introduced in definition 2.2.1, one can represent the infinitesimals (2.11) and (2.12) as sums of partial and functional derivatives of $\xi_{y(z)}\mathrm{d}z$, ξ_x, ξ_t, η_ϕ, cf. appendix A.

2.4 Infinitesimal generator, determining system of equations for the infinitesimals

Let $F([\overline{y(x')}], \overline{x}, \overline{t}, \underset{1}{\overline{\phi}}, \underset{2}{\overline{\phi}}, \underset{3}{\overline{\phi}}) = 0$ be a transformed FDE. Consider F as a function depending on the group parameter ε and expand F in a Taylor series (2.1) about $\varepsilon_0 = 0$, i.e. consider the equation

$$\left. F \right|_{\varepsilon=0} + \left.\frac{\partial F}{\partial \varepsilon}\right|_{\varepsilon=0} \varepsilon + \mathcal{O}(\varepsilon^2) = 0. \tag{2.13}$$

We calculate

$$\left.\frac{\partial F}{\partial \varepsilon}\right|_{\varepsilon=0} = \left[\int_G \xi_{y(x')}\mathrm{d}x'\frac{\partial}{\partial y(x')\mathrm{d}x'} + \xi_x\frac{\partial}{\partial x} + \xi_t\frac{\partial}{\partial t} + \eta_\phi\frac{\partial}{\partial \phi} \right.$$
$$\left. + \zeta_{;t}\frac{\partial}{\partial \phi_{,t}} + \int_G \zeta_{;y(x')}\mathrm{d}x'\frac{\partial}{\partial \phi_{,y(x')}\mathrm{d}x'} + \dots \right] F$$

and define

Definition 2.4.1 (Infinitesimal generator and its prolongation).

- The differential operator

$$X := \int_G \xi_{y(x')}\mathrm{d}x'\frac{\partial}{\partial y(x')\mathrm{d}x'} + \xi_x\frac{\partial}{\partial x} + \xi_t\frac{\partial}{\partial t} + \eta_\phi\frac{\partial}{\partial \phi}$$

  is called *infinitesimal generator* or simply *generator*.

- The differential operator

$$X^{(3)} := X + \zeta_{;t}\frac{\partial}{\partial \phi_{,t}} + \int_G \zeta_{;y(x')}\mathrm{d}x'\frac{\delta}{\delta\psi_{,y(x')}} + \zeta_{;tt}\frac{\partial}{\partial \phi_{,tt}} + \int_G \zeta_{;ty(x')}\mathrm{d}x'\frac{\delta}{\delta\phi_{,ty(x')}}$$
$$+ \int_G \zeta_{;xy(x')}\mathrm{d}x'\frac{\delta}{\delta\phi_{,xy(x')}} + \int_G\int_G \zeta_{;y(x'')y(x')}\mathrm{d}x'\mathrm{d}x''\frac{\delta}{\delta\phi_{,y(x'')y(x')}}$$
$$+ \zeta_{;ttt}\frac{\partial}{\partial \phi_{,ttt}} + \int_G \zeta_{;tty(x')}\mathrm{d}x'\frac{\delta}{\delta\phi_{,tty(x')}} + \int_G\int_G \zeta_{;ty(x'')y(x')}\mathrm{d}x'\mathrm{d}x''\frac{\delta}{\delta\phi_{,ty(x'')y(x')}}$$
$$+ \int_G \zeta_{;txy(x')}\mathrm{d}x'\frac{\delta}{\delta\phi_{,txy(x')}} + \int_G\int_G\int_G \zeta_{;y(x''')y(x'')y(x')}\mathrm{d}x'\mathrm{d}x''\mathrm{d}x'''\frac{\delta}{\delta\phi_{,y(x''')y(x'')y(x')}}$$
$$+ \int_G\int_G \zeta_{;y(x'')xy(x')}\mathrm{d}x'\mathrm{d}x''\frac{\delta}{\delta\phi_{,y(x'')xy(x')}} + \int_G\int_G \zeta_{;xy(x'')y(x')}\mathrm{d}x'\mathrm{d}x''\frac{\delta}{\delta\phi_{,xy(x'')y(x')}}$$
$$+ \int_G \zeta_{;xxy(x')}\mathrm{d}x'\frac{\delta}{\delta\phi_{,xxy(x')}}$$

  is called *prolongation* of X.

Using this definition and employing $F\big|_{\varepsilon=0} = F$, equation (2.13) reads

$$F + X^{(3)}F\varepsilon + \mathcal{O}(\varepsilon^2) = 0.$$

This equation is fulfilled in $\mathcal{O}(\varepsilon)$ if and only if

$$\left[X^{(3)}F\right]_{F=0} = 0. \tag{2.14}$$

Equation (2.14) constitutes an overdetermined system of linear FDE's for the infinitesimals $\xi_{y(z)}\mathrm{d}z$, ξ_x, ξ_t, η_ϕ. In order to formulate the system of equations, one has to insert the necessary infinitesimals, calculated in section 2.3, into equation (2.14). Since $\xi_{y(z)}\mathrm{d}z$, ξ_x, ξ_t, η_ϕ do not depend on derivatives of ϕ, all coefficients of all appearing derivatives of ϕ have to vanish which leads to a system of linear FDE's.

2.5 Global transformations

Knowing $\xi_{y(z)}\mathrm{d}z, \xi_x, \xi_t, \eta_\phi$, one obtains the global transformations using Lie's first theorem, cf. Ref. [14]:

Theorem 2.5.1 (Lie's first theorem). The global tansformation can be obtained by solving the following initial value problems:

$$\frac{\partial \overline{y(z)\mathrm{d}z}}{\partial \varepsilon} = \xi_{\overline{y(z)}\mathrm{d}z}, \qquad \frac{\partial \overline{x}}{\partial \varepsilon} = \xi_{\overline{x}}, \qquad \frac{\partial \overline{t}}{\partial \varepsilon} = \xi_{\overline{t}}, \qquad \frac{\partial \overline{\phi}}{\partial \varepsilon} = \eta_{\overline{\phi}}$$

with the initial values

$$\overline{y(z)\mathrm{d}z}(\varepsilon = 0) = y(z)\mathrm{d}z, \qquad \overline{x}(\varepsilon = 0) = x, \qquad \overline{t}(\varepsilon = 0) = t, \qquad \overline{\phi}(\varepsilon = 0) = \phi.$$

3 Lie symmetry analysis of the viscous Hopf-Burgers functional integro-differential equation

3.1 Three different approaches to Lie symmetry analysis

As already mentioned in the introduction, when working in the physical space, the main problem is to deal with the partial derivatives $\partial/\partial x$ and the transformation of the integration variable x. Different Lie symmetry analysis approaches for integro-differential equations have been proposed in literature, cf. Ref. [13]. In the following, we present three methods which could possibly be applied in our case of functional equations with functional derivatives.

1. We transform $y(z)$ instead of $y(z)dz$ and have to take into account the transformation of the integral term appearing in Hopf FDE's. In [14] and [15], IBRAGIMOV suggests to use the fact that X given by definition 2.4.1 is equivalent to a canonical Lie-Bäcklund operator $\tilde{X}$ which does not contain the term $\xi_x\partial/\partial x$. This implies $\tilde{X}$ is very suitable for the symmetry analysis of integro-differential equations. Hence, one might replace X by $\tilde{X}$ and perform the extended Lie symmetry analysis on functional integro-differential equations.

2. We transform $y(z)$ instead of $y(z)dz$ and consider the differential equation as an equation $F = 0$ where F depends on an integral term I and an integral-free term H, i.e. $F(H, I) = 0$. In order to get the correct determining system of equations for the infinitesimals, the transformation of F is expanded in a two-dimensional Taylor series about H and I. This method is presented by ZAWISTOWSKI in [17].

In [18] it is shown that using ZAWISTOWSKI's approach leads to results being equivalent to the results presented in IBRAGIMOV's study and that the Lie algebra of symmetry group transformations spanned by the infinitesimal generators containing integral terms is solvable, cf. Ref. [17] for the Vlasov-Maxwell integro-differential equation and [18] for the Benney integro-differential equations. ZAWISTOWSKI's approach leads to the following determining system of equations for the infinitesimals $\xi_{y(z)}dz$, ξ_x, ξ_t, η_ϕ:

$$\left[\mathscr{X}^{(3)}F - \int_G \frac{\partial \xi_{x'}}{\partial x'} f\, dx' \right]_{F=0} = 0.$$

One has to pay attention that an analogous formula for the transformed integral should be used during the calculation of $\zeta_{;t}$, $\zeta_{;xy(x)y(x)}$ and $\zeta_{;xxy(x)}$, cf. section 2.3, in order to determine the generator $\mathscr{X}^{(3)}$ given by definition 2.4.1, which makes this approach more complicated in our particular case of an equation with functional derivatives.

3. We transform $y(z)dz$ instead of $y(z)$ and introduce a transformation of x. Then, we perform the extended Lie symmetry analysis on the viscous Hopf-Burgers FDE (1.8). Presently, this approach is succesfully applied to (1.8) and we rediscover all known symmetries of the usual viscous Burgers equation (1.6). The results and a discussion comparing the symmetries of the viscous Hopf-Burgers FDE and the symmetries of the viscous Burgers equation are presented in the following subsections.

3.2 Local transformations of the viscous Hopf-Burgers functional integro-differential equation

3.2.1 Determining system of equations for the infinitesimals

In this section, we perform the extended Lie symmetry analysis on the viscous Hopf-Burgers FDE

$$F = \phi_{,t} - \int_G y(x)\left(i\phi_{,xy(x)y(x)} + \nu\phi_{,xxy(x)}\right)\,\mathrm{d}x = 0.$$

We consider the one-parameter Lie point transformations given by definition 2.1.2.

In order to calculate the determining system of equations for the infinitesimals $\xi_{y(z)}\mathrm{d}z$, ξ_x, ξ_t, η_ϕ, we consider equation (2.14):

$$\left[X^{(3)}F\right]_{F=0} = 0.$$

The generator $X^{(3)}$ is given by definition 2.4.1. The contributing summands of $X^{(3)}$ are given by

$$X^{(3)}_{\mathrm{modif}} = \int_G \xi_{y(x')}\mathrm{d}x'\,\frac{\partial}{\partial y(x')\mathrm{d}x'} + \zeta_{;t}\frac{\partial}{\partial\phi_{,t}} + \int_G\int_G \zeta_{;xy(x'')y(x')}\mathrm{d}x'\mathrm{d}x''\,\frac{\partial}{\partial\phi_{,xy(x'')y(x')}\mathrm{d}x'\mathrm{d}x''}$$
$$+ \int_G \zeta_{;xxy(x')}\mathrm{d}x'\,\frac{\partial}{\partial\phi_{,xxy(x')}\mathrm{d}x'}.$$

Applying $X^{(3)}_{\mathrm{modif}}$ to F, we get the equation

$$X^{(3)}_{\mathrm{modif}}F = \zeta_{;t} - \int_G \xi_{y(x)}\mathrm{d}x\left(i\phi_{,xy(x)y(x)} + \nu\phi_{,xxy(x)}\right) - \int_G y(x)\left(i\zeta_{;xy(x)y(x)}\mathrm{d}x + \nu\zeta_{;xxy(x)}\mathrm{d}x\right) = 0. \quad (3.1)$$

We insert the infinitesimals $\zeta_{;t}$ (cf. equation (2.10)), $\zeta_{;xy(x)y(x)}$ and $\zeta_{;xxy(x)}$ (cf. appendix A) and employ $F = 0$ in order to eliminate $\phi_{,t}$.

In the extended Lie symmetry analysis we deal with different types of derivatives. Inside the integral the functional derivatives $\delta/\delta y(x)$ are present. It was argued that these derivatives correspond to partial derivatives $\partial/\partial y_k$ in the discrete case, cf. section 2.1. Additionally, in (1.8), partial derivatives with respect to x are present, consider $\phi_{,xy(x)y(x)}$ and $\phi_{,xxy(x)}$, which do not have a correspondence in the discrete case. Also, different types of mixed functional-partial derivatives are present in the final formula (3.1). In the further procedure, in order to obtain proper symmetry transformations, for the mixed derivatives we have to assume that

- $\phi_{,y(x)}$ is not independent of $\phi_{,xy(x)}$,

- $\phi_{,ty(x)}$ is not independent of $\phi_{,txy(x)}$,

- $\phi_{,y(x)y(z)}$ is not independent of $\phi_{,xy(x)y(z)}$

etc. Hence, in the final step, we integrate by parts in order to remove the x-derivatives from the functional derivatives of ϕ. In order to do this, we have to eliminate the boundary integrals. One may choose between the following two options:

- The first option is to demand that $G \subseteq \mathbb{R}$ is bounded: $G = (a, b)$ with $-\infty < a < b < \infty$. If this holds true, one has to demand additionally that all appearing terms evaluated at $x = b$ equal to the same terms evaluated at $x = a$. Then all boundary integrals vanish. As this demands a huge number of restrictions, we do not choose G to be bounded. Instead, we choose the second option.

- The second option is to demand that $G \subseteq \mathbb{R}$ is not bounded. We restrict ourselves to the case $G = (-\infty, +\infty)$. If this holds true, one may impose the condition

$$U^t(x = \pm\infty) = 0. \tag{3.2}$$

As equation (1.5) states

$$\phi_{,y(x)} = i\langle U^t(x)e^{i(U^t,y)}\rangle,$$

we have

$$\phi_{,y(x)}\Big|_{x=\pm\infty} = 0.$$

Additionally, we impose that all functional derivatives of ϕ vanish for $x \to \pm\infty$, i.e.

$$\phi_{,ty(x)}\Big|_{x=\pm\infty} = 0, \qquad \phi_{,xy(x)}\Big|_{x=\pm\infty} = 0, \qquad \phi_{,y(x)y(x)}\Big|_{x=\pm\infty} = 0, \qquad \dots$$

Hence, all appearing boundary integrals vanish. For example,

$$\int_G \xi_{y(x)} \mathrm{d}x\, \phi_{,xy(x)y(x)} = -\int_G \frac{\partial \xi_{y(x)} \mathrm{d}x}{\partial x} \phi_{,y(x)y(x)} + \underbrace{\left[\xi_{y(x)}\phi_{,y(x)y(x)}\right]_{x=-\infty}^{x=+\infty}}_{=0}.$$

The resulting equation (3.1) has the form

$$
\begin{aligned}
0 = \mathscr{A} &+ \int_G \phi_{,y(z)}\mathscr{B}\, \mathrm{d}z + \int_G\int_G \phi_{,y(x)y(z)}\mathscr{C}\, \mathrm{d}z\mathrm{d}x + \int_G\int_G \phi_{,y(x)}\phi_{,y(z)}\mathscr{D}\, \mathrm{d}z\mathrm{d}x \\
&+ \int_G\int_G\int_G \phi_{,y(x)y(a)}\phi_{,y(z)}\mathscr{E}\, \mathrm{d}a\mathrm{d}z\mathrm{d}x + \int_G\int_G (\phi_{,y(x)})^2\phi_{,y(z)}\mathscr{F}\, \mathrm{d}z\mathrm{d}x \\
&+ \int_G \phi_{,ty(x)}\mathscr{G}\, \mathrm{d}x + \int_G\int_G (\phi_{,y(x)})^2\phi_{,y(z)y(z)}\mathscr{H}\, \mathrm{d}z\mathrm{d}x + \int_G \phi_{,y(x)}\phi_{,ty(x)}\mathscr{I}\, \mathrm{d}x \\
&+ \int_G \phi_{,ty(x)y(x)}\mathscr{J}\, \mathrm{d}x + \int_G\int_G \phi_{,y(z)y(x)y(x)}\mathscr{K}\, \mathrm{d}z\mathrm{d}x + \int_G\int_G \phi_{,y(x)y(x)}\phi_{,y(z)y(z)}\mathscr{L}\, \mathrm{d}z\mathrm{d}x.
\end{aligned}
$$

Since the infinitesimals $\xi_{y(z)}\mathrm{d}z$, ξ_x, ξ_t, η_ϕ do not depend on derivatives of ϕ, all coefficients of all appearing derivatives of ϕ have to vanish:

$$\mathscr{A} = \mathscr{B} = \mathscr{C} = \mathscr{D} = \mathscr{E} = \mathscr{F} = \mathscr{G} = \mathscr{H} = \mathscr{I} = \mathscr{J} = \mathscr{K} = \mathscr{L} = 0.$$

This leads to the following system of linear FDE's for the infinitesimals $\xi_{y(z)}\mathrm{d}z$, ξ_x, ξ_t, η_ϕ:

- $\mathscr{A} = 0$ reads

$$\frac{\partial \eta_\phi}{\partial t} - \int_G y(x)\left(i\frac{\partial^3 \eta_\phi}{\partial x \partial(y(x)\mathrm{d}x)^2} + v\frac{\partial^3 \eta_\phi}{\partial x^2 \partial y(x)\mathrm{d}x}\right)\mathrm{d}x = 0, \tag{3.3}$$

- $\mathscr{B} = 0$ reads

$$
\begin{aligned}
0 = {}& \nu\frac{\partial^2}{\partial z^2}\left(y(z)\frac{\partial\eta_\phi}{\partial\phi}\right) - \nu\frac{\partial^2}{\partial z^2}\left(y(z)\frac{\partial\xi_t}{\partial t}\right) - \frac{\partial\xi_{y(z)}}{\partial t} - \nu\frac{\partial^2\xi_{y(z)}}{\partial z^2} \\
& - 2iy(z)\frac{\partial^3\eta_\phi}{\partial z\partial y(z)\mathrm{d}z\partial\phi} + i\nu\int_G\frac{\partial^2}{\partial z^2}\left(y(x)y(z)\frac{\partial^3\xi_t}{\partial x\partial(y(x)\mathrm{d}x)^2}\right)\mathrm{d}x \\
& + i\int_G y(x)\frac{\partial^3\xi_{y(z)}}{\partial x\partial(y(x)\mathrm{d}x)^2}\,\mathrm{d}x - i\frac{\partial}{\partial z}\left(y(z)\frac{\partial^2\xi_z}{\partial z\partial y(z)\mathrm{d}z}\right) + 2i\frac{\partial}{\partial z}\left(y(z)\frac{\partial^2\eta_\phi}{\partial y(z)\mathrm{d}z\partial\phi}\right) \\
& - i\int_G\frac{\partial}{\partial x}\left(y(x)\frac{\partial^2\xi_{y(z)}}{\partial(y(x)\mathrm{d}x)^2}\right)\mathrm{d}x + i\frac{\partial^2}{\partial z^2}\left(y(z)\frac{\partial\xi_z}{\partial y(z)\mathrm{d}z}\right) \\
& + \nu^2\int_G\frac{\partial^2}{\partial z^2}\left(y(x)y(z)\frac{\partial^3\xi_t}{\partial x^2\partial y(x)\mathrm{d}x}\right)\mathrm{d}x + \nu\int_G y(x)\frac{\partial^3\xi_{y(z)}}{\partial x^2\partial y(x)\mathrm{d}x}\,\mathrm{d}x \\
& - 2\nu\int_G\frac{\partial}{\partial x}\left(y(x)\frac{\partial^2\xi_{y(z)}}{\partial x\partial y(x)\mathrm{d}x}\right)\mathrm{d}x - \nu\frac{\partial}{\partial z}\left(y(z)\frac{\partial^2\xi_z}{\partial z^2}\right) - \nu\frac{\partial^2}{\partial z^2}\left(y(z)\frac{\partial\eta_\phi}{\partial\phi}\right) \\
& + \nu\int_G\frac{\partial^2}{\partial x^2}\left(y(x)\frac{\partial\xi_{y(z)}}{\partial y(x)\mathrm{d}x}\right)\mathrm{d}x + 2\nu\frac{\partial^2}{\partial z^2}\left(y(z)\frac{\partial\xi_z}{\partial z}\right) + y'(z)\frac{\partial\xi_z}{\partial t} + y(z)\frac{\partial^2\xi_z}{\partial x\partial t} \\
& + i\frac{\partial^2}{\partial z^2}\left(y(z)\frac{\partial\xi_z}{\partial y(z)\mathrm{d}z}\right) - i\frac{\partial}{\partial z}\left(y(z)\frac{\partial^2\xi_z}{\partial z\partial y(z)\mathrm{d}z}\right) + i\int_G\frac{\partial^2}{\partial z\partial x}\left(y(z)y(x)\frac{\partial^2\xi_z}{\partial y(x)\mathrm{d}x\partial y(x)\mathrm{d}x}\right)\mathrm{d}x \\
& - i\int_G y(x)\frac{\partial}{\partial z}\left(y(z)\frac{\partial^3\xi_z}{\partial x(\partial y(x)\mathrm{d}x)^2}\right)\mathrm{d}x - \nu\int_G y'(z)y(x)\frac{\partial^3\xi_z}{\partial x^2\partial y(x)\mathrm{d}x}\,\mathrm{d}x \\
& - \nu\int_G y(z)y(x)\frac{\partial^4\xi_z}{\partial z\partial x^2\partial y(x)\mathrm{d}x}\,\mathrm{d}x - \nu\int_G\frac{\partial^3}{\partial x^2\partial z}\left(y(z)y(x)\frac{\delta\xi_z}{\delta y(x)}\right)\mathrm{d}x \\
& + 2\nu\int_G\frac{\partial^2}{\partial x\partial z}\left(y(z)y(x)\frac{\partial\xi_z}{\partial x\partial y(x)\mathrm{d}x}\right)\mathrm{d}x,
\end{aligned}
\tag{3.4}
$$

- $\mathscr{C} = 0$ reads

$$
\begin{aligned}
0 = {}& -i\frac{\partial}{\partial x}\left(y(x)\frac{\partial\eta_\phi}{\partial\phi}\right)\delta(x-z) + i\frac{\partial}{\partial x}\left(y(x)\frac{\partial\xi_t}{\partial t}\right)\delta(x-z) + i\frac{\partial\xi_{y(x)}}{\partial x}\delta(x-z) \\
& + \int_G\frac{\partial}{\partial z}\left(y(a)y(z)\frac{\partial^3\xi_t}{\partial a\partial(y(a)\mathrm{d}a)^2}\right)\delta(x-z)\,\mathrm{d}a + 2iy(x)\frac{\partial^2\xi_{y(z)}}{\partial x\partial y(x)\mathrm{d}x} \\
& - 2i\frac{\partial}{\partial x}\left(y(x)\frac{\partial\xi_{y(z)}}{\partial y(x)\mathrm{d}x}\right) + i\frac{\partial}{\partial x}\left(y(x)\frac{\partial\eta_\phi}{\partial\phi}\right)\delta(x-z) - i\frac{\partial}{\partial x}\left(y(x)\frac{\partial\xi_x}{\partial x}\right)\delta(x-z) \\
& - i\nu\int_G\frac{\partial}{\partial z}\left(y(a)y(z)\frac{\partial^3\xi_t}{\partial a^2\partial y(a)\mathrm{d}a}\right)\delta(x-z)\,\mathrm{d}a + \nu y(x)\frac{\partial^3\xi_{y(z)}}{\partial x^2\partial\phi} + \nu y(x)\frac{\partial^2\xi_{y(z)}}{\partial x^2} \\
& - \nu\frac{\partial}{\partial x}\left(y(x)\frac{\partial\xi_{y(z)}}{\partial x}\right) - \nu\frac{\partial}{\partial x}\left(y(x)\frac{\partial\xi_{y(z)}}{\partial x}\right) - 2iy(x)\frac{\partial}{\partial z}\left(y(z)\frac{\partial^2\xi_z}{\partial x\partial y(x)\mathrm{d}x}\right) \\
& + 2i\frac{\partial^2}{\partial x\partial z}\left(y(x)y(z)\frac{\partial\xi_z}{\partial y(x)\mathrm{d}x}\right),
\end{aligned}
\tag{3.5}
$$

- $\mathscr{D} = 0$ reads

$$
\begin{aligned}
0 = {}& -v^2 \frac{\partial^4}{\partial z^2 \partial x^2}\left(y(x)y(z)\frac{\partial \xi_t}{\partial \phi}\right) - v\frac{\partial^2}{\partial x^2}\left(y(x)\frac{\partial \xi_{y(z)}}{\partial \phi}\right) + 2iv\frac{\partial^2}{\partial z^2}\left(y(x)y(z)\frac{\partial^3 \xi_t}{\partial x \partial y(x)dx\partial \phi}\right) \\
& + 2iy(x)\frac{\partial^3 \xi_{y(z)}}{\partial x \partial y(x)dx\partial \phi} - i\frac{\partial}{\partial x}\left(y(x)\frac{\partial^2 \xi_x}{\partial x \partial \phi}\right)\delta(x-z) - 2iv\frac{\partial^3}{\partial z^2\partial x}\left(y(x)y(z)\frac{\partial^2 \xi_t}{\partial y(x)dx\partial \phi}\right) \\
& - 2i\frac{\partial}{\partial x}\left(y(x)\frac{\partial^2 \xi_{y(z)}}{\partial y(x)dx\partial \phi}\right)\delta(x-z) - 2i\frac{\partial}{\partial x}\left(y(x)\frac{\partial^2 \xi_{y(z)}}{\partial y(x)dx\partial \phi}\right) \\
& + 2i\frac{\partial}{\partial x}\left(y(x)\frac{\partial^2 \eta_\phi}{\partial \phi^2}\right)\delta(x-z) + i\frac{\partial^2}{\partial x^2}\left(y(x)\frac{\partial \xi_x}{\partial \phi}\right)\delta(x-z) + i\frac{\partial^2}{\partial x^2}\left(y(x)\frac{\partial \xi_x}{\partial \phi}\right)\delta(x-z) \\
& - v\frac{\partial}{\partial x}\left(y(x)\frac{\partial^2 \xi_{y(z)}}{\partial x\partial \phi}\right) - 2v\frac{\partial}{\partial x}\left(y(x)\frac{\partial^2 \xi_{y(z)}}{\partial x\partial \phi}\right) - v\frac{\partial}{\partial x}\left(y(x)\frac{\partial^2 \xi_{y(z)}}{\partial x\partial \phi}\right) \\
& + 2v\frac{\partial^2}{\partial x^2}\left(y(x)\frac{\partial \xi_{y(z)}}{\partial \phi}\right) + v^2\frac{\partial^4}{\partial z^2\partial x^2}\left(y(x)y(z)\frac{\partial \xi_t}{\partial \phi}\right) + v\frac{\partial^2}{\partial x^2}\left(y(x)\frac{\partial \xi_{y(z)}}{\partial \phi}\right) \\
& + v\frac{\partial^2}{\partial x^2}\left(y(x)\frac{\partial \xi_{y(z)}}{\partial \phi}\right) - 2v^2\frac{\partial^3}{\partial z^2\partial x}\left(y(x)y(z)\frac{\partial^2 \xi_t}{\partial x\partial \phi}\right) + v^2\frac{\partial^2}{\partial z^2}\left(y(x)y(z)\frac{\partial^3 \xi_t}{\partial x^2\partial \phi}\right) \\
& + i\frac{\partial^2}{\partial x^2}\left(y(x)\frac{\partial \xi_x}{\partial \phi}\delta(x-z)\right) + i\frac{\partial^2}{\partial x\partial z}\left(y(x)\frac{\partial \xi_x}{\partial \phi}\delta(x-z)\right) - i\frac{\partial}{\partial x}\left(y(x)\frac{\partial^2 \xi_x}{\partial \phi\partial x}\delta(x-z)\right) \\
& - i\frac{\partial}{\partial z}\left(y(x)y(z)\frac{\partial^3 \xi_z}{\partial x\partial \phi\delta y(x)}\right) + 2i\frac{\partial^2}{\partial z\partial x}\left(y(x)y(z)\frac{\partial^2 \xi_z}{\partial \phi\delta y(x)}\right)
\end{aligned}
\tag{3.6}
$$

- $\mathscr{E} = 0$ reads

$$
\begin{aligned}
0 = {}& 2iv\frac{\partial^3}{\partial z^2\partial x}\left(y(x)y(z)\frac{\partial \xi_t}{\partial \phi}\right)\delta(x-a) + i\frac{\partial}{\partial x}\left(y(x)\frac{\partial \xi_{y(z)}}{\partial \phi}\right)\delta(x-a) \\
& + 2\frac{\partial}{\partial x}\left(y(x)y(z)\frac{\partial^3 \xi_t}{\partial z\partial y(z)dz\partial \phi}\right)\delta(x-a) + 2iy(z)\frac{\partial^2 \xi_{y(x)}}{\partial z\partial \phi}\delta(a-z) + iy(x)\frac{\partial^2 \xi_{y(z)}}{\partial x\partial \phi}\delta(x-a) \\
& - 2\frac{\partial^2}{\partial z\partial x}\left(y(x)y(z)\frac{\partial^2 \xi_t}{\partial y(z)dz\partial \phi}\right)\delta(x-a) - 2i\frac{\partial}{\partial z}\left(y(z)\frac{\partial \xi_{y(x)}}{\partial \phi}\right)\delta(a-z) \\
& - i\frac{\partial}{\partial x}\left(y(x)\frac{\partial \xi_{y(z)}}{\partial \phi}\right)\delta(x-a) - 2i\frac{\partial}{\partial z}\left(y(z)\frac{\partial \xi_{y(x)}}{\partial \phi}\right)\delta(a-z) \\
& - iv\frac{\partial^3}{\partial z^2\partial x}\left(y(x)y(z)\frac{\partial \xi_t}{\partial \phi}\right)\delta(x-a) - i\frac{\partial}{\partial x}\left(y(x)\frac{\partial \xi_{y(z)}}{\partial \phi}\right)\delta(x-a) \\
& - iv\frac{\partial^3}{\partial z^2\partial x}\left(y(x)y(z)\frac{\partial \xi_t}{\partial \phi}\right)\delta(x-a) + iv\frac{\partial^2}{\partial z^2}\left(y(x)y(z)\frac{\partial^2 \xi_t}{\partial x\partial \phi}\right)\delta(x-a) \\
& + 2iv\frac{\partial^2}{\partial z\partial x}\left(y(x)y(z)\frac{\partial^2 \xi_t}{\partial z\partial \phi}\right)\delta(x-a) - 2iv\frac{\partial}{\partial x}\left(y(x)y(z)\frac{\partial^3 \xi_t}{\partial z^2\partial \phi}\right)\delta(x-a) \\
& + iy'(z)y'(x)\delta(a-z)\frac{\partial \xi_z}{\partial \phi},
\end{aligned}
\tag{3.7}
$$

- $\mathscr{F} = 0$ reads

$$
\begin{aligned}
0 = {}& iy(x)\frac{\partial^3 \xi_{y(z)}}{\partial x\partial \phi^2} - 2iv\frac{\partial^3}{\partial z^2\partial x}\left(y(x)y(z)\frac{\partial^2 \xi_t}{\partial \phi^2}\right) - i\frac{\partial}{\partial x}\left(y(x)\frac{\partial^2 \xi_{y(z)}}{\partial \phi^2}\right) \\
& - 2i\frac{\partial}{\partial x}\left(y(x)\frac{\partial^2 \xi_{y(z)}}{\partial \phi^2}\right) + iv\frac{\partial^2}{\partial z^2}\left(y(x)y(z)\frac{\partial^3 \xi_t}{\partial x\partial \phi^2}\right),
\end{aligned}
\tag{3.8}
$$

- $\mathcal{G} = 0$ reads

$$0 = 2iy(x)\frac{\partial^2 \xi_t}{\partial x \partial y(x)\mathrm{d}x} - 2i\frac{\partial}{\partial x}\left(y(x)\frac{\partial \xi_t}{\partial y(x)\mathrm{d}x}\right) - v\frac{\partial}{\partial x}\left(y(x)\frac{\partial \xi_t}{\partial x}\right) + vy(x)\frac{\partial^2 \xi_t}{\partial x^2}$$
$$- v\frac{\partial}{\partial x}\left(y(x)\frac{\partial \xi_t}{\partial x}\right), \tag{3.9}$$

- $\mathcal{H} = 0$ reads

$$0 = -2\frac{\partial^2}{\partial z \partial x}\left(y(x)y(z)\frac{\partial^2 \xi_t}{\partial \phi^2}\right) + \frac{\partial}{\partial z}\left(y(x)y(z)\frac{\partial^3 \xi_t}{\partial x \partial \phi^2}\right), \tag{3.10}$$

- $\mathcal{I} = 0$ reads

$$0 = -2i\frac{\partial}{\partial x}\left(y(x)\frac{\partial \xi_t}{\partial \phi}\right) - 2i\frac{\partial}{\partial x}\left(y(x)\frac{\partial \xi_t}{\partial \phi}\right) + 2iy(x)\frac{\partial^2 \xi_t}{\partial x \partial \phi}, \tag{3.11}$$

- $\mathcal{J} = 0$ reads

$$0 = iy(x)\frac{\partial \xi_t}{\partial x}, \tag{3.12}$$

- $\mathcal{K} = 0$ reads

$$0 = iy(x)\frac{\partial \xi_{y(z)}}{\partial x}, \tag{3.13}$$

- $\mathcal{L} = 0$ reads

$$0 = \frac{\partial^2}{\partial z \partial x}\left(y(x)y(z)\frac{\partial \xi_t}{\partial \phi}\right) - \frac{\partial^2}{\partial z \partial x}\left(y(x)y(z)\frac{\partial \xi_t}{\partial \phi}\right) - \frac{\partial}{\partial z}\left(y(x)y(z)\frac{\partial^2 \xi_t}{\partial x \partial \phi}\right). \tag{3.14}$$

3.2.2 Solution of the determining system of equations for the infinitesimals

- First of all, consider equation (3.13). Since this equation has to hold for all choices of $y \in L^2(G, \mathbb{R})$, the coefficient of y has to vanish and we get

$$\frac{\partial \xi_{y(z)}}{\partial x} = 0. \tag{3.15}$$

- Now, consider equation (3.12). Similarly, we get

$$\frac{\partial \xi_t}{\partial x} = 0. \tag{3.16}$$

- Then, consider equation (3.11). Similarly, we get

$$\frac{\partial \xi_t}{\partial \phi} = 0. \tag{3.17}$$

- Due to equation (3.17), equations (3.10) and (3.14) are fulfilled identically.

- Next, we consider equation (3.9). We apply the product rule and make use of equation (3.16). We get

$$\frac{\partial \xi_t}{\partial y(x)\mathrm{d}x} = 0. \tag{3.18}$$

Considering equations (3.16) and (3.17), we have $\xi_t = \xi_t(t)$.

- If we apply equations (3.17) and (3.15) to equation (3.7), we obtain

$$-4i\frac{\partial}{\partial z}\left(y(z)\frac{\partial \xi_{y(x)}}{\partial \phi}\right)\delta(a-z)-i\frac{\partial}{\partial x}\left(y(x)\frac{\partial \xi_{y(z)}}{\partial \phi}\right)\delta(x-a)+iy'(z)y'(x)\delta(a-z)\frac{\partial \xi_z}{\partial \phi} = 0. \tag{3.19}$$

Considering the case $a \neq z$ and taking into account that $y \in L^2(G,\mathbb{R})$ is an arbitrary function, we get

$$\frac{\partial \xi_{y(z)}}{\partial \phi} = 0. \tag{3.20}$$

With the above relation, equation (3.19) for $a \neq x$ leads to

$$\frac{\partial \xi_z}{\partial \phi} = 0. \tag{3.21}$$

which holds for each $z \in G$. If we substitute this result back into equation (3.19) and assume $x = z$, we find again the formula (3.20) which has to hold for each $z \in G$. As $y \in L^2(G,\mathbb{R})$ is an arbitrary function, equation (3.19) is fulfilled for $x \neq z$ as well.

- In virtue of equations (3.17) and (3.20), equation (3.8) is fulfilled identically.

- Now, we take a look at the remaining four equations (3.3) - (3.6). We start with equation (3.6). Considering equations (3.17), (3.20) and (3.21), equation (3.6) reads

$$2i\frac{\partial}{\partial x}\left(y(x)\frac{\partial^2 \eta_\phi}{\partial \phi^2}\right)\delta(x-z) = 0,$$

hence

$$\frac{\partial^2 \eta_\phi}{\partial \phi^2} = 0. \tag{3.22}$$

Equation (3.22) means there are functionals f, g such that

$$\eta_\phi = f([y(x')], t)\phi + g([y(x')], t). \tag{3.23}$$

For f we choose the ansatz

$$f([y(x')], t) = \int_G f_1(x', t)y(x')\mathrm{d}x' + f_2(t). \tag{3.24}$$

- The next equation we solve is equation (3.5). If we use equations (3.18) and (3.15) and apply the product rule, equation (3.5) reads

$$
\begin{aligned}
0 = {} & i\frac{\partial}{\partial x}\left(y(x)\frac{\partial\xi_t}{\partial t}\right)\delta(x-z) + i\frac{\partial\xi_{y(x)}}{\partial x}\delta(x-z) \\
& - i\frac{\partial}{\partial x}\left(y(x)\frac{\partial\xi_x}{\partial x}\right)\delta(x-z) - 2iy'(x)\frac{\partial\xi_{y(z)}}{\partial y(x)\mathrm{d}x} + 2iy'(z)y'(x)\frac{\partial\xi_z}{\partial y(x)\mathrm{d}x}.
\end{aligned}
\tag{3.25}
$$

Considering the case $x \neq z$ we get

$$
y'(z)\frac{\partial\xi_z}{\partial y(x)\mathrm{d}x} - \frac{\partial\xi_{y(z)}}{\partial y(x)\mathrm{d}x} = 0.
\tag{3.26}
$$

Although equation (3.26) allows a broader range of solutions, we restrict our considerations to the case

$$
\frac{\partial\xi_z}{\partial y(x)\mathrm{d}x} = 0, \quad \frac{\partial\xi_{y(z)}}{\partial y(x)\mathrm{d}x} = 0
\tag{3.27}
$$

and use the following ansätze for $\xi_{y(z)}\mathrm{d}z$ and ξ_z:

$$
\xi_{y(z)}\mathrm{d}z = c(z,t)\mathrm{d}z + c_0(z,t)y(z)\mathrm{d}z, \quad \xi_z = c_1(z,t) + c_2(z,t)y(z).
\tag{3.28}
$$

Next, we want to consider equation (3.25) without the restriction $x \neq z$, hence we integrate equation (3.25) with respect to $z \in G$. This leads to

$$
\frac{\partial}{\partial x}\left(y(x)\frac{\partial\xi_t}{\partial t}\right) + \frac{\partial\xi_{y(x)}}{\partial x} - \frac{\partial}{\partial x}\left(y(x)\frac{\partial\xi_x}{\partial x}\right) - 2\int_G y'(x)\frac{\partial\xi_{y(z)}}{\partial y(x)\mathrm{d}x}\,\mathrm{d}z + 2y'(x)\int_G y'(z)\frac{\partial\xi_z}{\partial y(x)\mathrm{d}x}\mathrm{d}z = 0.
$$

Now, we put in ansatz (3.28), make use of $\xi_t = \xi_t(t)$ and take into consideration that this equation has to hold for all choices of $y \in L^2(G,\mathbb{R})$, hence the coefficients of 1, y, y', y'', yy', $y'y'$ have to vanish:

$$
\frac{\partial c(x,t)}{\partial x} = 0 \Longrightarrow c = c(t),
\tag{3.29}
$$

$$
\frac{\partial c_0(x,t)}{\partial x} - \frac{\partial^2 c_1}{\partial x^2} = 0,
\tag{3.30}
$$

$$
\xi_t'(t) - \frac{\partial c_1}{\partial x} - c_0(x,t) = 0,
\tag{3.31}
$$

$$
c_2(x,t) = 0.
\tag{3.32}
$$

After differentiating equation (3.31) with respect to x we find

$$
\frac{\partial c_0}{\partial x} + \frac{\partial^2 c_1}{\partial x^2} = 0
$$

which, together with equation (3.30), gives

$$
\frac{\partial^2 c_1}{\partial x^2} = 0, \quad c_0 = c_0(t).
\tag{3.33}
$$

Considering this and equations (3.29) and (3.31), ansatz (3.28) reads

$$
\xi_z = c_3(t)z + c_4(t),
\tag{3.34}
$$

$$
\xi_{y(z)}\mathrm{d}z = c(t)\mathrm{d}z + \left(\xi_t'(t) - c_3(t)\right)y(z)\mathrm{d}z.
\tag{3.35}
$$

- Now, we are ready to deal with equation (3.4). If we use equations (3.35), (3.34), (3.24), (3.18) and (3.33), equation (3.4) reads

$$
\begin{aligned}
0 = {} & -\nu y''(z)\xi_t'(t) - c'(t) - \left(\xi_t''(t) - \frac{\partial^2 \xi_x}{\partial t \partial x}(t) \right) y(z) - \nu \left(\xi_t'(t) - \frac{\partial \xi_x}{\partial x}(t) \right) y''(z) \\
& - 2i y(z)\frac{\partial f_1(z,t)}{\partial z} + 2i\frac{\partial}{\partial z}(y(z)f_1(z,t)) + \nu \int_G y(x)\left(\xi_t'(t) - \frac{\partial \xi_z}{\partial z}(t) \right)\frac{\partial^2 \delta(x-z)}{\partial x^2}\,dx \\
& - 2\nu \int_G \frac{\partial}{\partial x}\left(y(x)\left(\xi_t'(t) - \frac{\partial \xi_x}{\partial x}(t) \right)\frac{\partial \delta(x-z)}{\partial x} \right)dx - \nu\frac{\partial}{\partial z}\left(y(z)\frac{\partial^2 \xi_z}{\partial z^2} \right) \\
& + \nu \int_G \frac{\partial^2}{\partial x^2}\left(y(x)\left(\xi_t'(t) - \frac{\partial \xi_x}{\partial x}(t) \right)\delta(x-z) \right)dx + 2\nu\frac{\partial^2}{\partial z^2}\left(y(z)\frac{\partial \xi_z}{\partial z}(t) \right) + y'(z)\frac{\partial \xi_z}{\partial t} \\
& + y(z)\frac{\partial^2 \xi_z}{\partial t \partial x}.
\end{aligned}
\tag{3.36}
$$

In this equation, the last two integrals involving the Dirac delta distribution vanish if we assume that

$$
y(x), y'(x) \longrightarrow 0 \text{ for } x \to \pm\infty.
$$

As equation (3.36) has to hold for all choices of $y \in L^2(G, \mathbb{R})$, the coefficients of 1, y, y' and y'' have to vanish. We evaluate the coefficient of y'' in $x = z$ and get

$$
-c'(t) = 0 \Longrightarrow c(t) = const. =: c \in \mathbb{R},
\tag{3.37}
$$

$$
-\xi_t''(t) + 2\frac{\partial^2 \xi_z}{\partial t \partial z}(t) = 0,
\tag{3.38}
$$

$$
2i f_1(z,t) + \frac{\partial \xi_z}{\partial t} = 0,
\tag{3.39}
$$

$$
2\nu\frac{\partial \xi_z}{\partial z}(t) - \nu\xi_t'(t) = 0.
\tag{3.40}
$$

From the above system we first use the relations (3.40) and (3.23), take into account (3.33) and substitute this into equation (3.3) to get

$$
\frac{1}{2}i\frac{\partial^2 \xi_z}{\partial t^2}\int y(x)dx\,\phi + \frac{\partial g}{\partial t} - \int_G y(x)\left(i\frac{\partial^3 g}{\partial x \partial(y(x)dx)^2} + \nu\frac{\partial^3 g}{\partial x^2 \partial y(x)dx} \right)dx = 0.
\tag{3.41}
$$

This equation has to hold for every ϕ, hence the coefficient of ϕ has to vanish. Together with (3.33) this furnishes

$$
\frac{\partial^2 \xi_z}{\partial t^2} = 0,
\tag{3.42}
$$

hence equation (3.41) reads

$$
\frac{\partial g}{\partial t} - \int_G y(x)\left(i\frac{\partial^3 g}{\partial x \partial(y(x)dx)^2} + \nu\frac{\partial^3 g}{\partial x^2 \partial y(x)dx} \right)dx = 0,
\tag{3.43}
$$

i.e. g has to fulfill the viscous Hopf-Burgers FDE. This is expected by the classical Lie symmetry analysis as the considered differential equation is linear; we see that this result is furnished by the extended Lie symmetry analysis as well. With (3.42), the system (3.38) - (3.40) has the solution

$$
\xi_t = 2a_1 t + a_2 + a_4 t^2,
\tag{3.44}
$$

$$
\xi_x = a_1 x + a_3 + t x a_4 + a_5 t,
\tag{3.45}
$$

$$
f_1 = \frac{1}{2}ixa_1 + \frac{1}{2}ia_3 + a_6,
\tag{3.46}
$$

where $a_1, a_2, a_3, a_4, a_5, a_6 \in \mathbb{R}$.

We insert equations (3.45), (3.44) and (3.37) into equation (3.35) to get

$$\xi_{y(z)}\mathrm{d}z = (a_1 + ta_4)y(z)\mathrm{d}z + a_7\mathrm{d}z,$$

with $a_7 \in \mathbb{R}$.

Theorem 3.2.1 (Local transformations of the viscous Hopf-Burgers FDE). The infinitesimals of the viscous Hopf-Burgers FDE are given by

$$\xi_t = 2a_1 t + a_2 + a_4 t^2,$$
$$\xi_x = a_1 x + a_3 + txa_4 + a_5 t,$$
$$\xi_{y(z)}\mathrm{d}z = (a_1 + ta_4)y(z)\mathrm{d}z + a_6\mathrm{d}z,$$
$$\eta_\phi = \left(\frac{1}{2}i\int (a_4 x + a_5)y(x)\mathrm{d}x + a_7\right)\phi + g([y(x')], t),$$

where $a_1, a_2, a_3, a_4, a_5, a_6, a_7 \in \mathbb{R}$ are arbitrary constants and g is an arbitrary functional which has to fulfill the viscous Hopf-Burgers FDE.

The associated generators read

$$X_1 = x\frac{\partial}{\partial x} + 2t\frac{\partial}{\partial t} + \int_G y(x)\mathrm{d}x\frac{\partial}{\partial y(x)\mathrm{d}x},$$

$$X_2 = \frac{\partial}{\partial t},$$

$$X_3 = \frac{\partial}{\partial x},$$

$$X_4 = 2tx\frac{\partial}{\partial x} + 2t^2\frac{\partial}{\partial x} + 2t\int_G y(x)\mathrm{d}x\frac{\partial}{\partial y(x)\mathrm{d}x} + i\int xy(x)\mathrm{d}x\phi\frac{\partial}{\partial\phi},$$

$$X_5 = 2t\frac{\partial}{\partial x} + i\int y(x)\mathrm{d}x\phi\frac{\partial}{\partial\phi}$$

$$X_6 = \int_G \mathrm{d}x\frac{\partial}{\partial y(x)\mathrm{d}x},$$

$$X_7 = \phi\frac{\partial}{\partial\phi},$$

$$X_8 = g([y(x')], t)\frac{\partial}{\partial\phi}.$$

3.3 Symmetry breaking restrictions

The Lie symmetry analysis furnishes symmetries of the viscous Hopf-Burgers FDE without respecting physical restrictions. If we incorporate such physical restrictions, we lose some of the calculated symmetries which the viscous Hopf-Burgers FDE exhibits considered as a mathematical equation detached from any physical conditions. The loss of symmetries by incorporating physical restrictions is called *symmetry breaking*. This section is devoted to restrictions on ϕ breaking some of the calculated symmetries $X_1, \ldots, X_g$.

In [6], HOPF states conditions which have to be fulfilled by Hopf functionals

$$\phi([y(x)], t) = \left\langle e^{i(U^t, y)} \right\rangle = \int_{L^2(G, \mathbb{R}^3)} e^{i(v, y)} f^t([v(x)]) \, d[v(x)],$$

cf. equation (1.3). These conditions may be derived from conditions which are imposed on the associated probability density functional f^t. The definition of a probability density functional requires f^t to fulfill the following two conditions:

Definition 3.3.1 (Probability density functional). f^t is called a *probability density functional* if and only if

1. f^t is real-valued and non-negative, i.e.

$$f^t([v(x)]) \in \mathbb{R}_0^+. \tag{3.47}$$

2. The integral of f^t over the whole domain of integration equals 1, i.e.

$$\int_{L^2(G, \mathbb{R}^3)} f^t([v(x)]) \, d[v(x)] = 1. \tag{3.48}$$

As in this paper we restrict ourselves to the one-dimensional case and make use of the viscous Burgers equation instead of the incompressible Navier-Stokes equations, solutions of the viscous Hopf-Burgers FDE do not have to fulfill any conditions related to incompressibility. There remain three conditions which a solution ϕ of the viscous Hopf-Burgers FDE has to fulfill. We define:

Definition 3.3.2 (Physically relevant solution of the viscous Hopf-Burgers FDE). Let ϕ be a solution of the viscous Hopf-Burgers FDE. ϕ is a *physically relevant solution* of the viscous Hopf-Burgers FDE if and only if

1. $\phi^*([y(x)], t) = \phi([-y(x)], t)$ where ϕ^* denotes the complex conjugate of ϕ,

2. $\phi(0, t) = 1$,

3. $|\phi([y(x)], t)| \leq 1$.

These three conditions are implied by restriction (3.47) and equation (3.48).

In subsection 3.2.2 we showed that the extended Lie symmetry analysis furnishes eight generators X_1, $X_2, X_3, X_4, X_5, X_6, X_7, X_g$ as the local transformations depend on seven parameters $a_1, a_2, a_3, a_4, a_5, a_6, a_7$ and on a functional g. Especially, the generators X_7 and X_g associated with symmetries of ϕ are independent. As the conditions given by definition 3.3.2 do not influence any symmetries corresponding to transformations of the independent variables $([y(x')], x, t)$, the generators associated with transformations of the independent variables are not changed if we are looking for physically relevant solutions: We have

$$X_i^{\text{phys}} = X_i, \qquad i \in \{1, 2, 3, 4, 5, 6\}.$$

Thus, it suffices to have a look at X_7 and X_g. If ϕ shall be a physically relevant solution, X_7 and X_g are not independent. In order to see this, decompose g in

$$X_g = g([y(x')], t) \frac{\partial}{\partial \phi}$$

into a constant part $g_1 \in \mathbb{C}$ and a non-constant part $g_2 = g_2([y(x')], t)$, i.e.

$$g([y(x')], t) = g_1 + g_2([y(x')], t).$$

Here, g_2 is a solution of the viscous Hopf-Burgers FDE, however, it is not a characteristic functional, i.e. the conditions given by definition 3.3.2 must be fulfilled for the transformed functional $\overline{\phi}$ but not for g_2 separately. We get the decomposition $X_g = X_{g_1} + X_{g_2}$ with

$$X_{g_1} := g_1 \frac{\partial}{\partial \phi}, \qquad g_1 \in \mathbb{C},$$

$$X_{g_2} := g_2([y(x')], t) \frac{\partial}{\partial \phi}.$$

We replace X_7 and X_g by the two generators

$$X_7 + X_{g_1} = (\phi + g_1) \frac{\partial}{\partial \phi}, \qquad f, g_1 \in \mathbb{C},$$

$$X_{g_2} = g_2([y(x')], t) \frac{\partial}{\partial \phi}$$

and calculate the associated global transformations by solving the Lie initial value problems

$$\frac{\partial \overline{\phi}}{\partial \varepsilon} = \overline{\phi} + g_1, \qquad\qquad \frac{\partial \overline{\phi}}{\partial \varepsilon} = g_2([y(x')], t),$$

$$\overline{\phi}(\varepsilon = 0) = \phi, \qquad\qquad \overline{\phi}(\varepsilon = 0) = \phi.$$

The solutions are given by

$$\overline{\phi}([y(x)], t) = \phi([y(x)], t)e^{\varepsilon} + (e^{\varepsilon} - 1)g_1, \qquad \varepsilon \in \mathbb{R}, \quad g_1 \in \mathbb{C}, \qquad (3.49)$$

$$\overline{\phi}([y(x)], t) = \phi([y(x)], t) + g_2([y(x)], t)\varepsilon, \qquad \varepsilon \in \mathbb{R}. \qquad (3.50)$$

In the following, we investigate the consequences of the conditions given by definition 3.3.2 for $\varepsilon \in \mathbb{R}$ and $g_1 \in \mathbb{C}$ if $\overline{\phi}$ is given by equation (3.49).

1. As ϕ has to fulfill $\phi^*([y(x)], t) = \phi([-y(x)], t)$, we have

$$g_1 \in \mathbb{R}.$$

2. As ϕ has to fulfill $\phi^*([y(x)], t) = \phi([-y(x)], t)$ and $\phi(0, t) = 1$ and since $g_1 \in \mathbb{R}$, we get

$$g_1 = -1.$$

This shows that X_7 and X_g are not independent.

3. As ϕ has to fulfill $|\phi([y(x)], t)| \leq 1$, we have

$$\varepsilon \in (-\infty, 0]$$

and the generator $X_7 + X_{g_1}$ associated with the physically relevant symmetry reads

$$X_7^{\text{phys}} := (X_7 + X_{g_1})^{\text{phys}} = (\phi - 1) \frac{\partial}{\partial \phi}.$$

Next, we investigate the consequences of the conditions given by definition 3.3.2 for $\varepsilon \in \mathbb{R}$ and $g_2 = g_2([y(x)], t)$ if $\overline{\phi}$ is given by equation (3.50).

1. As ϕ has to fulfill condition $\phi^*([y(x)], t) = \phi([-y(x)], t)$, we have

$$g_2^*([y(x)], t) = g_2([-y(x)], t). \tag{3.51}$$

2. As ϕ has to fulfill $\phi^*([y(x)], t) = \phi([-y(x)], t)$ and $\phi(0, t) = 1$, using equation (3.51) we get

$$g_2(0, t) = 0. \tag{3.52}$$

3. As ϕ has to fulfill condition $|\phi([y(x)], t)| \leq 1$, we have

$$\mathrm{Re}(\phi)\mathrm{Re}(g_2\varepsilon) + \mathrm{Im}(\phi)\mathrm{Im}(g_2\varepsilon) \leq 0. \tag{3.53}$$

Altogether, the generator X_{g_2} associated with the physically relevant symmetry reads

$$X_g^{\mathrm{phys}} := X_{g_2}^{\mathrm{phys}} = g_2([y(x')], t)\frac{\partial}{\partial \phi}$$

where g_2 fulfills conditions (3.51) - (3.53).

In the end of this section, we want to compare the calculated physically relevant symmetries with the symmetries of the viscous Burgers equation, cf. Ref. [14].

Symmetries of the viscous Hopf-Burgers FDE	_Symmetries of the viscous Burgers equation_
$X_1^{\mathrm{phys}} = 2t\frac{\partial}{\partial t} + x\frac{\partial}{\partial x} + \int_G y(x)\mathrm{d}x\frac{\partial}{\partial y(x)\mathrm{d}x},$	$\mathbb{X}_1 = 2t\frac{\partial}{\partial t} + x\frac{\partial}{\partial x} - U\frac{\partial}{\partial U},$
$X_2^{\mathrm{phys}} = \frac{\partial}{\partial t},$	$\mathbb{X}_2 = \frac{\partial}{\partial t},$
$X_3^{\mathrm{phys}} = \frac{\partial}{\partial x},$	$\mathbb{X}_3 = \frac{\partial}{\partial x},$
$X_4^{\mathrm{phys}} = 2tx\frac{\partial}{\partial x} + 2t^2\frac{\partial}{\partial x}$ $+ 2t\int_G y(x)\mathrm{d}x\frac{\partial}{\partial y(r)\mathrm{d}x} + i\int xy(x)\mathrm{d}x\phi\frac{\partial}{\partial \phi},$	$\mathbb{X}_4 = 2t^2\frac{\partial}{\partial t} + 2tx\frac{\partial}{\partial x} + (x - 2tU)\frac{\partial}{\partial U},$
$X_5^{\mathrm{phys}} = 2t\frac{\partial}{\partial x} + i\int y(x)\mathrm{d}x\phi\frac{\partial}{\partial \phi},$	$\mathbb{X}_5 = 2t\frac{\partial}{\partial x} + \frac{\partial}{\partial U}.$
$X_6^{\mathrm{phys}} = \int_G \mathrm{d}x\frac{\partial}{\partial y(x)\mathrm{d}x},$	
$X_7^{\mathrm{phys}} = (\phi - 1)\frac{\partial}{\partial \phi},$	
$X_g^{\mathrm{phys}} = g_2([y(x')], t)\frac{\partial}{\partial \phi}.$	

Table 3.1.: Comparison between the symmetries of the viscous Hopf-Burgers FDE and the viscous Burgers equation.

Here, g_2 is a solution of the viscous Hopf-Burgers FDE satisfying the three conditions (3.51) - (3.53). For X_7^{phys}, the group parameter ε is restricted to be non-positive, i.e. $\varepsilon \in (-\infty, 0]$. For all the other generators we have $\varepsilon \in \mathbb{R}$.

We see that we rediscover the analog forms of the symmetries of the viscous Burgers equation generated by $\mathbb{X}_1$ (scaling), $\mathbb{X}_2$ (time translation), $\mathbb{X}_3$ (space translation), $\mathbb{X}_4$ and $\mathbb{X}_5$ (Galilei invariance).

Generator	Global transformations associated with the generator			
X_1^{phys}	$\bar{t} = te^{2\varepsilon},$	$\bar{x} = xe^{\varepsilon},$	$\overline{y(x)\mathrm{d}x} = y(x)\mathrm{d}x\,e^{\varepsilon},$	$\bar{\phi} = \phi,$
X_2^{phys}	$\bar{t} = t + \varepsilon,$	$\bar{x} = x,$	$\overline{y(x)\mathrm{d}x} = y(x)\mathrm{d}x,$	$\bar{\phi} = \phi,$
X_3^{phys}	$\bar{t} = t,$	$\bar{x} = x + \varepsilon,$	$\overline{y(x)\mathrm{d}x} = y(x)\mathrm{d}x,$	$\bar{\phi} = \phi,$
X_4^{phys}	$\bar{t} = \frac{t}{1-2t\varepsilon},$	$\bar{x} = \frac{x}{(1-2t\varepsilon)}$	$\overline{y(x)\mathrm{d}x} = \frac{y(x)\mathrm{d}x}{1-2t\varepsilon},$	$\bar{\phi} = \phi \exp\left(i\frac{\varepsilon}{1-2t\varepsilon}\int_G xy(x)\mathrm{d}x\right),$
X_5^{phys}	$\bar{t} = t,$	$\bar{x} = x + 2t\varepsilon,$	$\overline{y(x)\mathrm{d}x} = y(x)\mathrm{d}x$	$\bar{\phi} = \phi \exp\left(i\varepsilon \int_G y(x)\mathrm{d}x\right),$
X_6^{phys}	$\bar{t} = t,$	$\bar{x} = x,$	$\overline{y(x)\mathrm{d}x} = y(x)\mathrm{d}x + \varepsilon\mathrm{d}x,$	$\bar{\phi} = \phi,$
X_7^{phys}	$\bar{t} = t,$	$\bar{x} = x,$	$\overline{y(x)\mathrm{d}x} = y(x)\mathrm{d}x,$	$\bar{\phi} = \phi e^{\varepsilon} + (1 - e^{\varepsilon}),$
X_8^{phys}	$\bar{t} = t,$	$\bar{x} = x,$	$\overline{y(x)\mathrm{d}x} = y(x)\mathrm{d}x,$	$\bar{\phi} = \phi + g([y(x')], t)\varepsilon.$

Table 3.2.: Global transformations of the viscous Hopf-Burgers FDE.

4 Conclusions

This paper continues the work of OBERLACK and WACŁAWCZYK, cf. Ref. [1] and [2], where the classical Lie symmetry analysis is extended from partial differential equations to equations with functional derivatives and performed in the Fourier space. Here, we introduce the procedure of applying the extended Lie symmetry analysis in the *physical* space. This corresponds to the case when both functional derivatives and spatial derivates with respect to the integration variable are present in the functional integro-differential equation. The method is based on the transformation of the product $y(x)\mathrm{d}x$ appearing in the integral term.

As an example, we consider the *viscous Hopf-Burgers functional integro-differential equation*, i.e. the functional formulation of the viscous Burgers equation. We perform the extended Lie symmetry analysis on the viscous Hopf-Burgers FDE to find the eight symmetries given in table 3.1. Furthermore, we take a brief look at symmetry breaking restrictions and indicate physically relevant symmetries, i.e. symmetries such that $\bar{\phi}$ fulfills the conditions for the characteristic functional. We see that only *statistical* symmetries, i.e. symmetries associated with transformations of the dependent variable ϕ, are influenced if we demand $\bar{\phi}$ to be a characteristic functional. The construction of physically relevant invariant solutions remains a task for future work. Finally, we compare the symmetries of the viscous Hopf-Burgers FDE with the symmetries of the viscous Burgers equation: We are able to rediscover all the five symmetries.

The most significant result of this paper consists in demonstrating that the extended Lie symmetry analysis works for the considered functional equation and that it is not only able to rediscover symmetries of the considered equation but also to furnish new, unknown symmetries associated with the Hopf formulation of the viscous Burgers equation having purely statistical origin. The presented extension of the Lie symmetry analysis can be a useful tool for the analysis of FDE's containing functional derivatives.

Presently, the underlying equation is the viscous Burgers equation. For future work one might consider the Hopf-Navier-Stokes FDE and perform the extended Lie symmetry analysis on this equation to determine the moments of the solutions of the Navier-Stokes equations (see also [19]). Furthermore, one might choose more sophisticated ansätze during the solution procedure of the determining system of equations for the infinitesimals. Hopefully, less restrictive ansätze will lead to further new symmetries.

We believe that the presented machinery is highly relevant to a variety of important functional differential equations and functional integro-differential equations in physics, especially in continuum mechanics. As the numerical treatment of FDE's is difficult because of the high dimensionality and since very little is known on how to treat and solve FDE's analytically, the presented methods may give a chance to treat equations which so far have been put aside because of the missing analytical methods. Additionally, unknown symmetries may be discovered which would be pleasant since symmetries illuminate the properties of the physical model equations.

5 Acknowledgments and contributions

The authors are thankful to Wolfgang Kollmann for his useful comments and discussions concerning the paper.

M. Oberlack and M. Wacławczyk contributed analysis methods; D. D. Janocha and M. Wacławczyk solved the determining system of equations for the infinitesimals; D. D. Janocha wrote the paper within the scope of his master's thesis.

Bibliography

[1] Oberlack, M.; Wacławczyk, M.: On the extension of Lie group analysis to functional differential equations. *Arch. Mech.* **2006**, *58*, 597-618

[2] Wacławczyk, M.; Oberlack, M.: Application of the extended Lie group analysis to the Hopf functional formulation of the Burgers equation. *J. Math. Phys.* **2013**, *54*, 072901

[3] Oberlack, M., Rosteck, A.: New statistical symmetries of the multi-point equations and its importance for turbulent scaling laws. *Disc. and Cont. Dyn. Sys., Ser. S* **2010**, *3*, 451-471

[4] Wacławczyk, M.; Staffolani, N.; Oberlack, M.; Rosteck, A.; Wilczek, M.; Friedrich, R.: Statistical Symmetries of the Lundgren-Monin-Novikov Hierarchy. *Phys. Rev. E* **2014**, *90*, 013022

[5] Lundgren, T. S.: Distribution functions in the statistical theory of turbulence. *Phys. Fluids*, **1967**, *10*, 969-975

[6] Hopf, E.: Statistical Hydromechanics and Functional Calculus. *J. Rational Mech. Anal.* **1952**, *1*, 87-123

[7] Hosokawa, I.: Monin-Lundgren hierarchy versus the Hopf equation in the statistical theory of turbulence. *Phys. Rev. E,* **2006**, *73*, 067301

[8] Hosokawa, I.; Yamamoto K.: Numerical Study of the Burgers' Model of Turbulence Based on the Characteristic Functional Formalism. *Phys. Fluids* **1970**, *13*, 1683-1692

[9] Hosokawa, I.; Yamamoto K.: Energy decay of Burgers' model of turbulence. *Phys. Fluids* **1976**, *19*, 1423-1424

[10] Grebenev, N. N.; Nazarenko, S. V.; Medvedev S. B.; Schwab I. V.; Chirkunov, Yu A.: Self-similar solution in the Leith model of turbulence: anomalous power law and asymptotic analysis. *J. Phys. A: Math. Theoret.* **2014**, *47*, 025501

[11] Alt, H. W.: Lineare Funktionalanalysis. Springer-Verlag, 2012

[12] Gelfand, I. M.; Fomin, S. W.: Calculus of variations. *Prentice Hall* **1963**, New Jersey

[13] Ibragimov, N. H.; Kovalev, V. F.; Pustovalov V. V.: Symmetries of integro-differential equations: a survey of methods illustrated by the Benny equations. *Nonl. Dyn.* **2002**, *28*, 135-153

[14] Ibragimov, N. H.: CRC Handbook of Lie Group Analysis of Differential Equations, Vol. 1: Symmetries, Exact Solutions and Conservation Laws. *CRC Press* **1994**

[15] Ibragimov, N. H.: CRC Handbook of Lie Group Analysis of Differential Equations, Vol. 3: New Trends in Theoretical Developments and Computational Methods. *CRC Press* **1996**

[16] Klauder, J. R.: A Modern Approach to Functional Integration, Birkhäuser **2011**

[17] Zawistowski, Z. J.: Symmetries of Integro-Differential Equations. *Rep. Math. Phys.* **2001**, *48*, 269-275

[18] Özer, T.: Symmetry group analysis of Benney system and an application for shallow-water equations, *Mech. Res. Commun.* **2005** , *32*, 241-254

[19] Oberlack, M.; Wacławczyk, M.; Rosteck, A.; Avsarkisov, V.: Symmetries and their importance for statistical turbulence theory, *Mechanical Engineering Reviews*, **2015**, *2*, 15-00157

A Infinitesimals of the dependent variables

For the viscous Hopf-Burgers FDE we need the three infinitesimals $\zeta_{;t}$, $\zeta_{;xy(x)y(x)}$ and $\zeta_{;xxy(x)}$. $\zeta_{;t}$ is given by equation (2.10):

$$\zeta_{;t} = \frac{\partial \eta_\phi}{\partial t} + \phi_{,t}\left(\frac{\partial \eta_\phi}{\partial \phi} - \frac{\partial \xi_t}{\partial t}\right) - (\phi_{,t})^2 \frac{\partial \xi_t}{\partial \phi} - \int_G \phi_{,y(x')}\frac{\partial \xi_{y(x')}\mathrm{d}x'}{\partial t} - \int_G \phi_{,y(x')}\phi_{,t}\frac{\partial \xi_{y(x')}\mathrm{d}x'}{\partial \phi}$$
$$- \int_G \phi_{,x'y(x')}\frac{\partial \xi_{x'}}{\partial t}y(x')\mathrm{d}x' - \int_G \phi_{,x'y(x')}\phi_{,t}\frac{\partial \xi_{x'}}{\partial \phi}y(x')\mathrm{d}x'.$$

$\zeta_{;xy(x)y(x)}$ and $\zeta_{;xxy(x)}$ are given below:

$$\zeta_{;xy(x)y(x)} = \frac{\partial^3 \eta_\phi}{\partial x \partial (y(x)\mathrm{d}x)^2} + 2\phi_{,y(x)}\frac{\partial^3 \eta_\phi}{\partial x \partial y(x)\mathrm{d}x \partial \phi} - \phi_{,t}\frac{\partial^3 \xi_t}{\partial x \partial (y(x)\mathrm{d}x)^2}$$
$$- 2\phi_{,t}\phi_{,y(x)}\frac{\partial^3 \xi_t}{\partial x \partial y(x)\mathrm{d}x \partial \phi} - \int_G \phi_{,y(z)}\frac{\partial^3 \xi_{y(z)}\mathrm{d}z}{\partial x \partial (y(x)\mathrm{d}x)^2}$$
$$- 2\int_G \phi_{,y(z)}\phi_{,y(x)}\frac{\partial^3 \xi_{y(z)}\mathrm{d}z}{\partial x \partial y(x)\mathrm{d}x \partial \phi} + (\phi_{,y(x)})^2\frac{\partial^3 \eta_\phi}{\partial x \partial \phi^2} - \phi_{,t}(\phi_{,y(x)})^2\frac{\partial^3 \xi_t}{\partial x \partial \phi^2}$$
$$- \int_G \phi_{,y(z)}(\phi_{,y(x)})^2\frac{\partial^3 \xi_{y(z)}\mathrm{d}z}{\partial x \partial \phi^2} - 2\phi_{,ty(x)}\frac{\partial^2 \xi_t}{\partial x \partial y(x)\mathrm{d}x} - 2\phi_{,y(x)}\phi_{,ty(x)}\frac{\partial^2 \xi_t}{\partial x \partial \phi}$$

$$-2\int_G \phi_{,y(z)y(x)}\frac{\partial^2\xi_{y(z)}\mathrm{d}z}{\partial x\partial y(x)\mathrm{d}x} - 2\int_G \phi_{,y(x)}\phi_{,y(z)y(x)}\frac{\partial^2\xi_{y(z)}\mathrm{d}z}{\partial x\partial\phi} - \phi_{,xy(x)}\frac{\partial^2\xi_x}{\partial x\partial y(x)\mathrm{d}x}$$

$$-\phi_{,y(x)}\phi_{,xy(x)}\frac{\partial^2\xi_x}{\partial x\partial\phi} + \phi_{,y(x)y(x)}\frac{\partial^2\eta_\phi}{\partial x\partial\phi} - \phi_{,t}\phi_{,y(x)y(x)}\frac{\partial^2\xi_t}{\partial x\partial\phi}$$

$$-\int_G \phi_{,y(z)}\phi_{,y(x)y(x)}\frac{\partial^2\xi_{y(z)}\mathrm{d}z}{\partial x\partial\phi}$$

$$+2\phi_{,xy(x)}\frac{\partial^2\eta_\phi}{\partial y(x)\mathrm{d}x\partial\phi} - 2\phi_{,t}\phi_{,xy(x)}\frac{\partial^2\xi_t}{\partial y(x)\mathrm{d}x\partial\phi} - \int_G \phi_{,xy(z)}\frac{\partial^2\xi_{y(z)}\mathrm{d}z}{\partial(y(x)\mathrm{d}x)^2}$$

$$-2\int_G \phi_{,xy(z)}\phi_{,y(x)}\frac{\partial^2\xi_{y(z)}\mathrm{d}z}{\partial y(x)\mathrm{d}x\partial\phi} - 2\int_G \phi_{,y(z)}\phi_{,xy(x)}\frac{\partial^2\xi_{y(z)}\mathrm{d}z}{\partial y(x)\mathrm{d}x\partial\phi}$$

$$+2\phi_{,y(x)}\phi_{,xy(x)}\frac{\partial^2\eta_\phi}{\partial\phi^2} - 2\phi_{,t}\phi_{,y(x)}\phi_{,xy(x)}\frac{\partial^2\xi_t}{\partial\phi^2} - \int_G \phi_{,xy(z)}(\phi_{,y(x)})^2\frac{\partial^2\xi_{y(z)}\mathrm{d}z}{\partial\phi^2}$$

$$-2\int_G \phi_{,y(z)}\phi_{,y(x)}\phi_{,xy(x)}\frac{\partial^2\xi_{y(z)}\mathrm{d}z}{\partial\phi^2} - 2\phi_{,xy(x)}\phi_{,ty(x)}\frac{\partial\xi_t}{\partial\phi}$$

$$-2\int_G \phi_{,xy(x)}\phi_{,y(z)y(x)}\frac{\partial\xi_{y(z)}\mathrm{d}z}{\partial\phi} - (\phi_{,xy(x)})^2\frac{\partial\xi_x}{\partial\phi} - \int_G \phi_{,xy(z)}\phi_{,y(x)y(x)}\frac{\partial\xi_{y(z)}\mathrm{d}z}{\partial\phi}$$

$$-2\phi_{,xty(x)}\frac{\partial\xi_t}{\partial y(x)\mathrm{d}x} - 2\phi_{,y(x)}\phi_{,xty(x)}\frac{\partial\xi_t}{\partial\phi} - \phi_{,xxy(x)}\frac{\partial\xi_x}{\partial y(x)\mathrm{d}x} - \phi_{,y(x)}\phi_{,xxy(x)}\frac{\partial\xi_x}{\partial\phi}$$

$$-2\int_G \phi_{,xy(z)y(x)}\frac{\partial\xi_{y(z)}\mathrm{d}z}{\partial y(x)\mathrm{d}x} - 2\int_G \phi_{,y(x)}\phi_{,xy(z)y(x)}\frac{\partial\xi_{y(z)}\mathrm{d}z}{\partial\phi} + \phi_{,xy(x)y(x)}\frac{\partial\eta_\phi}{\partial\phi}$$

$$-\phi_{,t}\phi_{,xy(x)y(x)}\frac{\partial\xi_t}{\partial\phi} - \int_G \phi_{,y(z)}\phi_{,xy(x)y(x)}\frac{\partial\xi_{y(z)}\mathrm{d}z}{\partial\phi} - \phi_{,ty(x)y(x)}\frac{\partial\xi_t}{\partial x}$$

$$-\int_G \phi_{,y(z)y(x)y(x)}\frac{\partial\xi_{y(z)}\mathrm{d}z}{\partial x} - \phi_{,xy(x)y(x)}\frac{\partial\xi_x}{\partial x} - \phi_{,xxy(x)}\frac{\partial\xi_x}{\partial y(x)\mathrm{d}x}$$

$$-\phi_{,xy(x)}\frac{\partial^2\xi_x}{\partial x\partial y(x)\mathrm{d}x} - \phi_{,xxy(x)}\phi_{,y(x)}\frac{\partial\xi_x}{\partial\phi} - \phi_{,xy(x)}\phi_{,xy(x)}\frac{\partial\xi_x}{\partial\phi} - \phi_{,xy(x)}\phi_{,y(x)}\frac{\partial^2\xi_x}{\partial\phi\partial x}$$

$$-\int_G \phi_{,y(x)}\phi_{,zy(z)}y(z)\frac{\partial^3\xi_z}{\partial x\partial\phi\partial y(x)\mathrm{d}x}\mathrm{d}z - 2\int_G \phi_{,zy(x)y(z)}y(z)\frac{\partial^2\xi_z}{\partial x\partial\partial y(x)\mathrm{d}x}\mathrm{d}z$$

$$-\int_G \phi_{,zxy(z)}y(z)\frac{\partial^2\xi_z}{(\partial y(x)\mathrm{d}x)^2}\mathrm{d}z - \int_G \phi_{,zy(z)}y(z)\frac{\partial^3\xi_z}{\partial x(\partial y(x)\mathrm{d}x)^2}\mathrm{d}z$$

$$-\int_G \phi_{,xy(x)}\phi_{,zy(z)}y(z)\frac{\partial^2\xi_z}{\partial\phi\partial y(x)\mathrm{d}x}\mathrm{d}z - \int_G \phi_{,y(x)}\phi_{,zxy(z)}y(z)\frac{\partial^2\xi_z}{\partial\phi\partial y(x)\mathrm{d}x}\mathrm{d}z$$

$$-2\int_G \phi_{,zxy(x)y(z)}y(z)\frac{\partial\xi_z}{\partial y(x)\mathrm{d}x}\mathrm{d}z - 2\int_G \phi_{,zxy(x)y(z)}\phi_{,y(x)}y(z)\frac{\partial\xi_z}{\partial\phi}\mathrm{d}z$$

$$-2\int_G \phi_{,zy(x)y(z)}\phi_{,xy(x)}y(z)\frac{\partial\xi_z}{\partial\phi}\mathrm{d}z - \int_G \phi_{,xy(x)y(x)}\phi_{,zy(z)}y(z)\frac{\partial\xi_z}{\partial\phi}\mathrm{d}z$$

BEI GRIN MACHT SICH IHR WISSEN BEZAHLT

- Wir veröffentlichen Ihre Hausarbeit, Bachelor- und Masterarbeit

- Ihr eigenes eBook und Buch - weltweit in allen wichtigen Shops

- Verdienen Sie an jedem Verkauf

Jetzt bei www.GRIN.com hochladen und kostenlos publizieren